About the Author

Rev. Robert Zaw Lwin, M.S., is a native of Myanmar (Burma). He belongs to the Chin ethnic group who live in the west central lowlands – Magway (west central), Bago (south central coast) and Rakhine (central west coast). His home diocese is Pyay (previously Prome) which was established by the American La Salette Missionaries (1937-1976).

He completed his studies for a Bachelor's Degree in Philosophy (Major) and Psychology (Minor) from the University of La Salette, Santiago City, Isabella, Philippines in 2002. He earned a Baccalaureate in Sacred Theology (S.T.B) from the Loyola School of Theology and a Master of Arts, Major in Pastoral Ministry (M.P.M), from the Ateneo De Manila University, Manila, Philippines in 2009.

Beyond that, he completed his La Salette Seminary formation in the La Salette Province of the Philippines and was ordained to the Priesthood on October 12, 2009.

Before coming to the United States, he served as a Pastor of a parish in the Archdiocese of Mandalay, in central Myanmar. He has served as an Associate Pastor at the Catholic Church of St. Ann, Marietta, Georgia. He graduated with a Doctor of Ministry Degree (D.Min.) from the Catholic University of America, Washington, D.C., on May 15, 2021.

He and the growing number of native La Salette Missionaries now serving in Myanmar are truly grateful to the North American La Salettes who courageously established the La Salette ministry of reconciliation in his homeland and do their part to make Mary's message known to all her people.

Dedication

This book is dedicated to those who work tirelessly to conserve and protect the natural environment. I remember my grandparents, parents, and teachers who taught me to conserve energy and resources and to love and take care of God's creation.

Everyone who loves and preserves wildlife sanctuary is loved by all creatures of God, let alone being loved by the Creator God. May everyone learn to use our natural resources wisely and help protect Mother Nature to the best of their abilities.

Fr. Robert Salai Zaw Lwin, M.S., D.Min.

A view from space of the continent of Africa

Rev. Robert Salai Zaw Lwin, M.S., D.Min.

Conversion, Creation, and the Common Good: a La Salette Perspective

Imprimi Potest:

Rev. Fr. William Kaliyadan, M.S., Provincial Superior, Missionaries of Our Lady of La Salette, Province of Mary, Mother of the Americas, 85B New Park Avenue, Hartford, CT 06106-2184 USA

Printed in the United States of America

ISBN: 978-1-946956-53-8

Booklet Design, Editing, and Digital Formatting: David Laferriere and Fr. Ron Gagne, M.S.

Table of Contents

Introduction

There is no more significant sign than the sign of love that those who love us show us daily, even by their mere presence. If even the presence of humans, our loved ones, and friends make a big difference in our lives, how much more will the visit of the Mother of God do? We always need answers to our unending questions in life, such as "Why did the Blessed Mother visit La Salette, France? Why should God be concerned with our lives, from creation to cremation? What do God's warnings and the good news of His blessings have to do with us?

To answer these questions, we shall explore and explain conversion, creation, care for creation, and the common good. We will use the teachings of the Church, the works of theologians, spiritual writers, and reliable authors on spirituality, creation, conversion, the common good, and the care for creation. My personal experiences and those of my friends and loved ones shall be included to present the theoretical and pastoral views on the themes we will discuss.

Our Lady of La Salette speaks with Melanie and Maximin on the Holy Mountain in France

We will also see and understand how God prepares the whole creation before creating humans in His image and likeness. God prepares the way for the coming of Jesus, His Son, into the world through natural and supernatural signs to make us aware of our need for conversion, to care for creation, and to work for the common good.

Chapter #1: explores and discusses the impact of human indifference on creation and the cry of creation for help. It also clarifies how God responds to its cry through a Marian apparition and invites humans to conversion. Humans must cooperate with God and collaborate in being open to our conversion of mind and for creation. To provide their every need, God, out of His perpetual love, created and prepared creation for His beloved people (Psalm 105:1-45).

Eva Segovia paints four beautiful symbols of areas of ecology: Human DNA, our Animals, the Earth's elements, and our Plants

Chapter #2: focuses on the need for human conversion to combat the environmental crisis. here are signs that highlight the human responsibility to care for creation. We must all learn to see, understand, appreciate, and care for God's creation. Our human ignorance, negligence, and failures necessitate our continual conversion. This is truer for believers, Christians, and all people of goodwill.

Conversion for Christians is as indispensable as breath for the sustenance of human life. As we all perish when we stop breathing, our Christian life dies when we stop being converted and care for creation.

Chapter #3: presents the necessity of human conversion and the role of Sacred Scripture in human life. The Scriptures and the Church's teaching guide us to care for God's creation. Our human dignity and identity are linked with the rest of creation. Even many Catholics, let alone other Christians, believers, and people of goodwill, sometimes fail to see this connection between humans and creation. To care for creation is to care for ourselves.

Chapter #4: presents how frail we humans are and how we always need redirection. Our every effort to reform our lives has life-changing effects on everyone and the whole of creation. The realities around us and the Sacred Scriptures greatly help our struggle to become better or even the best persons we are called to become. Indeed, conversion is our daily Christian responsibility so that we will make choices and decisions for the good and well-being of the whole world.

Chapter #5: helps us understand that our capacity to care for, work for, and live for the common good is at the core of our existence. Our concern and care for the common good constitute our identity as Catholics, Christians, and people of goodwill. The sense of the common good helps us realize our duty and responsibility to care for God's creation.

Realizing our human and Christian duty and responsibility towards creation makes us aware of its needs. This awareness then moves us into action, especially in using Artificial Intelligence responsibly to preserve, protect, and care for creation.

Chapter One

Reflecting on a Change of Heart

"A clean heart create for me, God; renew within me a steadfast spirit."
(Psalm 51:12)

Lord, make me realize that
I always need You in my daily life.
You are the only best source and resource
of every good thing.
Continue to appoint saints and angels
as well as sages and experts
to journey with me along the path to conversion
until I am finally united with you forever.

Our Human Environment

We sometimes live in environments where we grow into the person God wants us to be and are challenged to experience God's continuing love as we struggle to feel God's presence in real-life circumstances and situations. Either way, we need to respond to God's call which sometimes requires us to change our way of life. No matter how challenging or complex life may be, God is always ready to bring joy into our lives. And there is no greater joy than knowing and experiencing God's love in concrete circumstances and situations.

The Catholic Church has experienced, known, and taught the love of God that people can experience through the five senses. More realistic is the encounter with the heavenly realities, especially the encounter with Jesus, the Incarnate Son of God, through the visionaries of Marian apparitions and in the sacramentals and the sacraments. This reality of encounter with God through His agents and representatives conveys the message of God that humans must not ignore but to which we must pay attention.

During the apparitions of the Blessed Mother, she always reminds as well as warns people to listen to God and reform their lives so that God's blessings will be at work in their lives. People must do something to respond to God's invitation and avoid all the things that endanger and will do further damage to their dignity and identity. The renewal of their lives will affect the whole of God's creation.

Our need to be converted

Human conversion is the divine remedy for the healing of creation and the foundation of the common good. Instinctively, humans seek and follow both natural and supernatural signs to make life better day after day. We humans are gifted with and, therefore, tasked to find ways and means to address life's issues that affect us personally and communally. We are often challenged to investigate the conditions of our environment and repair the damage done to it or change anything that is not conducive to our livelihood, health, and life itself. We sometimes must undergo conversion to uplift the plight of creation and to work for the common good, which are our inalienable duties as humans.

Conversion, care for creation, and work for the common good are indispensable tasks for Christians. Conversion is a lifetime struggle and necessity for all Christians, although some may experience it in drastic ways.

Two signs of our urgent and expected response to the cry of creation are our sensitivity and immediate actions to address the needs of creation. It is compulsory for Christians to realize the divine remedy for creations' ailing conditions. Despite the need, one thing that has become very prevalent in human history is human indifference to God's call. Human indifference to God's call (Matthew 11:17) is never new but always present in every generation.

Learning to listen to the voice of God

Being indifferent to God and His creation calls for our urgent response. Such a response involves a radical change of heart and lifestyle. In the process of change, three essential things will help us overcome human indifference: conversion, care for God's creation, and the common good. These have always been overlooked and neglected despite the Church's teachings and warnings for many generations.

We can achieve these things only if we constantly listen to the voice of God. Why listen to the divine voice constantly reminding us to reform our lives? During His time, Jesus challenged His listeners to be attentive to God's voice so that they would be sensitive to the needs of others. Jesus remarked concerning the indifference of the people of the time, and He gave testimony to John the Baptist, who was the voice in the desert, calling on God's people to return to God. The role of John the Baptist was to prepare for the coming of Jesus. (1)

Living in an age of indifference

To awaken people from their indifference to God's invitation to His presence in others and the whole creation, Jesus openly said, "To what shall I compare this generation? It is like children who are in marketplaces and call to one another, 'We played the flute for you, but you did not dance; we sang a dirge, but you did not mourn.'" (2)

We may wonder why many people did not hear and accept Jesus's message. It could be their indifference to Jesus's very presence and invitation. In this age of human indifference, how shall we become aware of our indifference to one another and God? How far can human indifference affect our relationship with God, others, and creation?

What is it that we lose when we become indifferent towards life? Where is human indifference rooted? Matthew Kelly, in his book Life is Messy, notes that "the opposite of love is indifference. The Pied Piper leads millions into despair...but it can only be expressed by one person to another person, from one group of people to another group of people. Indifference cannot survive in this world outside the human heart. (3)

We explore in this book how human indifference can affect and eventually destroy all lives and the importance of conversion and work for the common good. Disregarding God's continuing warnings against human stubbornness, many in our world continue to ignore, neglect, and forget about the real dangers we all face.

We may not realize that all dangers have their root causes. Some dangers have roots in inordinate human desires, reckless behaviors, and lifestyles. The only thing that helps us to overcome human indifference is the love of God. The God who Himself is love asks us to live in His love and spread it everywhere. (4)

Our God, who has everlasting love for all his children and the whole of creation, uses both natural and supernatural signs to warn us against dangers. God uses these signs to warm our estranged hearts and to mend our broken hearts and even shattered lives for some so that we may continue to enjoy the beautiful life He has given us and the beautiful world we all live in. Can we know, feel, encounter, and witness His loving presence and continuing warnings in real-life circumstances and situations?

To find answers to these questions, we turn to Church teachings which are ever new, beautiful, and realistic. The beauty of the Catholic faith is that it reveals our ability to experience God in and through His creation, let alone through prayers and the sacraments.

The call to conversion

As God sent angels, prophets, sages, and leaders in the past, He sent into the world and our lives His Son (5) whom Mary bore when the time was right (6). Historical facts reveal that since 40 A.D. (7), the Blessed Mother has appeared to many visionaries to bring God's message to all of God's people.

The Creator's call for converting His children, as made known to two children, Maximin and Melanie, at La Salette, France, on September 19, 1846, is essential to combat climate change and care for creation. God sent the Blessed Mother to warn us against human indifference. Despite God's continuing warnings through both natural and supernatural signs, Christians and the people of goodwill, let alone ordinary people, often miss the connection between conversion, creation, and the common good. One of the reasons why we miss the warning signs is that we fail to realize and admit that beliefs, customs, and cultural traits can reinforce or hide our awareness of ecological crises.

Forming a creation-centered attitude

Based on the Christian perspective and understanding of conversion, creation, and the common good, let us ask the following questions:

- What are the constitutive elements of conversion, caring for creation, and working for the common good?

- What are the signs we always see around us that remind us of the need for our own conversion and our responsibility to care for creation and work for the common good?

There are many signs pointing to everything that has been happening in the world and the impending dangers we will all have to face unless we change our way of life.

Aside from natural signs, supernatural signs are revealed to us through creation. In truth and reality, creation provides natural signs that point to supernatural ones. In the history of the Catholic Church, we read that many supernatural signs have reminded or warned people of impending dangers in every generation.

The apparitions of the Blessed Virgin Mary, the Mother of Jesus Christ, are some of the supernatural signs that we Catholic Christians believe and witness to in every generation. One of the modern Marian apparitions, the apparition of the Blessed Virgin Mary at La Salette and her message of forgiveness and reconciliation, is explained well in my first book, A La Salette Spirituality of Forgiveness and Reconciliation for Everyday Living (September 19, 2021). (8)

Our Blessed Mother always relays the message of God, our Loving Father and Creator, regarding human needs, some of which are our conversion and responsibility to care for creation and work for the common good.

At La Salette, the Blessed Mother had a great message for God's children. The message touches on humanity's need for conversion, the care they must provide for creation, and working for the common good. Consequently, everyone will benefit from the wonders and richness of creation as nature regenerates itself and produces bountiful fruits and harvests.

The Blessed Mother said, "If [my people] are converted, the stones and rocks will be changed into mounds of wheat, and the potatoes will be self-sown in the fields." (9)

Are Marian apparitions real?

Many may ask, "Are Marian Apparitions real?" Aside from their queries about supernatural phenomena, many people of different beliefs and cultures argue and seek proofs for phenomena like Marian apparitions. From the standpoint of the Catholic Church, Marian apparitions are real because many saints and sages of every age have attested to their reality.

A German depiction of Our Lady of La Salette with the two children

One of the great contemplatives and spiritual writers, M. Basil Pennington, in his book *Mary Today: The Challenging Woman, Model for Maturing Christians*, observes that "the first reports of Marian visitations are woven into the rich fabric of early Christian times when historical fact and legend came together to create the clothing of the faith. These reports did not see the need to distinguish between truth and legend so long as they served the true faith. They were, in time, enshrined in the prayer life, both liturgical and devotional, of the Christian people. (10)

The truths about Marian apparitions may only be relevant for some believers, except for some who acknowledge the reality of Marian apparitions. For Catholics, Marian apparitions are real and pertinent to their beliefs. The Blessed Mother's apparitions and the power of her intercession before God are real for those who believe in the Trinitarian God.

A Buddhist woman and Mary's prayers

When I was a pastor of a mission parish as a newly ordained La Salette Missionary priest in a staunch Buddhist region in Myanmar (Burma), I witnessed a Buddhist woman bringing flowers to the grotto of the Blessed Mother. When I asked her why she regularly brought flowers to the Blessed Mother, although she is a Buddhist, to my surprise, she replied, "I believe in her presence, her love, and her intercession for me and my family. I pray that the mother asks God to heal my sick husband."

I was awestruck by her belief in the power of intercessory prayers, especially those of the Blessed Mother. There is a valuable lesson we can learn from the belief of this Buddhist woman, especially about the role of the Blessed Virgin Mary, who has appeared to God's beloved children in very generation.

Since even non-Catholics can feel and experience the loving presence of the Mother of God, we can say that "so prominent a feature of Catholic life has [the] apparitions become that the expression, 'Marian Age,' is often assigned to our times." (11)

What's an apparition anyway?

Despite things happening according to their natural process, every Marian apparition "tells with new freshness the awe-inspiring stories of the visits of the Mother of God to her children on earth." (12) We might sometimes wonder what the term "apparition" really might mean! Therefore, I believe that it is necessary to refresh our knowledge and understanding of the term "apparition" to understand better the meaning of the message the Blessed Mother gives to visionaries:

"Apparition is the name assigned to certain kinds of supernatural visions, either bodily or visible, and is the term most frequently applied to the various visions of Our Lady. The teaching of the Church regarding apparitions is comparable to her teaching about miracles which frequently stem from these apparitions." (13)

Some questions about apparitions

Why does the Blessed Mother always appear only to certain simple people, not exclusively to devout Catholics, let alone to all? Why must God send the Blessed Mother to make His message known to all? Does God not have any other ways of letting us know His will and plans or of warning us against the dangers encroaching upon us? To answer these questions, let's look into the trends in human society with which we all have to wrestle.

I concur that popular cultures are powerful, and the privileged who embrace and espouse such cultures have more rights and resources than anyone else. However, God always surprises us, especially primarily through miraculous and supernatural events involving the simple and/or the forgotten of human society.

Marian apparitions are supernational events that happen during natural day-to-day life. The difference between natural and supernatural events can be distinguished using human faculties and the light of faith. Natural events and supernatural events are distinct and yet linked together. The former often meet human exceptions, while the latter seldom do. Regardless of human expectations, God uses natural signs and events to show us supernatural signs and events.

The ordinary can become quite extraordinary

Unlike common human expectations that are colored by popular cultures, God sends the Blessed Mother instead to remind or warn His people of impending dangers. The visionaries of many Marian apparitions tell us that the Blessed Mother usually smiled at them. On the contrary, Pennington notices, "at La Salette, the woman who usually smiled at her children wept and wept before she smiled" (14). Pennington contends that

"most fundamental is Mary's loving care. She is still a Jewish mother in Canaan who wants to take care of things and her children. She is the woman who stood at the cross, knows what her Son has suffered, and knows her motherly role. The response to such love must take the form of reparation. In some way, we must make up for what our sin has caused, the sorrow and pain in the hearts of Jesus and Mary." (15)

How shall we reform our lives to console the weeping Mother of God and God's groaning creation? From a spiritual standpoint, faith and prayer are the most effective weapons against spiritual and moral dangers. They can help us see the needs of others and those of God's creation.

Reflection Questions:

1. Who is a person whom I admire for their motivation in life? What example in their life comes to mind?
2. What event in my personal or family life caused me to change for the better? What have I learned from that event?
3. How well do you respond to the Church's yearly call to conversion in the Lenten season? Do you "give up" something or do something special for others?

Chapter Two

Exploring Respect for God's Creation

Care for God's creation is good for the soul

A spiritual practice of conversion, the restoration and care for God's creation, and the work for the common good will help us live our human and Christian identity to the fullest. At the outset, it is necessary as an initial step to investigate and understand the meaning of spirituality in general, particularly of Christian spirituality. More importantly, Marian spirituality may have profound connections to the care for God's creation because it is generally maternal love that sees and cares for the neediest of human families and neighborhoods.

As we delve into the sources of Marian spirituality, we shall learn about scholars' thoughts, ideas, and understanding of spirituality. We may also discover the real meanings of their concepts and ideas about spirituality. The scholars' expertise can help us practice our chosen spiritual practices daily.

We can do it more effectively with proper understanding because we joyfully and consistently apply only what we truly know and understand. The Christian understanding of spirituality will be our starting point as we venture into the wealth of spiritual practices. The unique way Christians and Christian scholars articulate their experiences and practices of their chosen spiritual practices is outstanding for me personally.

Spirituality is a transcendental and deep experience with a supernatural being or a deity. It can be defined as a person's transcendental or supernatural experience of God through a life of devotion, prayer, and religious practices.

Patrick J. Hartin, in his book Exploring the Spirituality of the Gospels, provides the meaning and application of Christian spirituality in daily living. He notes, "Originally, the term spirituality was used in a religious and Christian context to refer to the mystical encounter between the individual and God that was expressed in a devotional life of prayer and pious practices." (16) Even if one has a devotion to any deity and does corresponding practices to earn blessings or forgiveness from that deity, how deeply that devotion affects the devotee can make a wonderful difference. For Catholics, devotional life is very important in living our Catholic spirituality. It is a simple and concrete way of communicating with God.

According to Hartin, things are different when we discuss Christian spirituality because "the pulsating heart of Christian spirituality is Jesus Christ. The message and person of Christ animates the existence of Christians and gives them meaning and direction through the power of the Spirit." (17) Christians believe, understand, and practice their faith and the corresponding teachings in unique ways that have their foundations in Jesus Christ.

Adherents of other religious traditions may have different outlooks on spiritual practices and understandings of being spiritual. For Christians, being spiritual means participating in the life of Christ. Spirituality is rooted in Christ, without whom life has no direction and meaning.

Human life, Christians firmly believe, is rooted in the life of the Risen Christ, which sustains every living being both on earth and in Heaven. For Christians, every human activity must reflect God's love to the extent that no vices have a significant place in their lives. Every fiber of their being and every single act manifests the living God they believe in.

Their lives are rooted in and sustained by the very life of their God, which is seen and understood in loving relationships with others, the whole of creation, and with Himself. Christian spirituality, which is different from other spiritualities, is one of the main points in their lives. The experts in Christian spirituality have a lot to teach us about the essence of Christian spirituality.

Learning more about our spirituality

Several authors' works can help us learn and understand Christian spirituality. For instance, Jordan Aumann, in his book Spiritual Theology, defines Christian spirituality as the way Christians participate in Christ's mystery and other virtuous acts.

Jesus, human and divine, gives the same life He shares with the Father and the Holy Spirit. Such a life that Jesus shares with Christians is an impressive expression and inexhaustible reality of the Trinitarian God. Christ's revelation and sharing of His life in communion with the Father and the Holy Spirit constitutes the essence of Christian spirituality, mainly of a Trinitarian nature. (18)

Jesus' life and ministry constantly challenge Christians of every generation to live their lives differently from people of different religious or faith traditions. Every Christian responds to Jesus' bidding to follow Him closely so that they may learn to love every creature and create the reality of God daily. Jesus calls His followers to be down to earth and love Him. It means they must love Him in concrete situations as life's circumstances call for everywhere and in every generation.

Philip Sheldrake construes that "each Christian, or group of Christians, is confronted by the gospel message from the past, which nevertheless demands a response within the concrete circumstances of today." (19) Christians must remember as well as urge others to acknowledge that every single human activity has an impact, sometimes even an irreversible impact, on the lives of all creatures and all created realities. They must all come together to act on changes in human activities as a response to God's call for cooperation with His grace and collaboration with one another to restore the original state of God's creation.

Lessons for a teenager

I remember an elderly neighbor chastising me when I was just a teenager on one rainy day because I was building a little water dam in the rain that blocked the torrent from flowing smoothly. As a child, I found it fun, but she reprimanded me to stop it because the larger torrential flood could happen to the point of affecting the basement of her house, her plants, and some crops. Only now have I realized that even just one little activity can lead to the destruction of the possessions of others.

We can only imagine how devasting it would be to the natural environment if we continue building water dams that can hold tons of water without properly considering the safety of the local people, their homes, and their livelihoods. I remember the dam that the Chinese businesspeople tried to build in Myitsone, Myitkyina, Myanmar. Many locals were forcefully removed from their homes and relocated to places where they could not make a living.

They had become internally displaced people (IDP) long before the civil unrest and revolution in Myanmar in 2021. The dam projects that required cutting trees and reshaping the landscapes destroyed their homes and farms. Domestic animals lost their pastures, while many wild animals, birds, and insects lost their habitats. Such reckless human activities damaged every part of creation. Devasting erosions also occurred because of mining that helped the corporations find precious stones such as jade, ruby, and emeralds.

Other natural resources, such as teak wood, different trees, geckoes, snakes, and countless wild animals, were taken from the locals. Before the dam project, many locals never destroyed the soil and lands where they planted their crops and raised their farm animals.

On the contrary, the mining corporations had no concern for the safety and lives of humans, let alone other creatures. Even without the dam built upstream of the Ayeyarwady River in northern Myanmar, there must have been many different ways to use irrigation systems for agriculture and advancement in human society without any environmental detriment. Environmentalists, local communities, the local Catholic churches, and the people of goodwill in Myanmar have been at the forefront of protecting the environment and locals.

Left: Artist's conception of the (presently postponed) Myitsone Dam to be built by Chinese workmen in Myitkyina, Myanmar, despite the protestations of countless native ecology-minded Burmese; right: The Ayeyawady River before the dam was to be completed upstream.

Natural disasters are signs of creation's anguish and anger toward everything that causes damage to nature's life. On July 1, 2024, the whole city of Myitkyina was flooded, surpassing the danger level scientists and locals know and have experienced.

Having lived in complex political and economic situations for more than three years, the locals had minimal resources they could use for their safety. Without electricity and an internet connection, the meager help the locals wanted to provide the flood victims was ineffective. The experts believed that the leading cause of the flood was the melting ice on Mount Putao. If the dam had been built and broken, the disaster would have been worse than it was.

It was not only the capital of Myitkyina but also the Ayeyarwaddy River, the delta region of Myanmar, which was affected since the flood water flowed through the delta into the Bay of Bengal. The Ayeyarwaddy River, as Myanmar's sole source of agriculture, needs a lot of protection and care to provide water and other life-giving resources for humans and natural habitats.

We can only imagine the devastating consequences of both human and natural disasters. I am sure that very few would have thought that climate change could affect even Myanmar. However, it has unfortunately affected the agricultural nation of Myanmar under the rule of incompetent and selfish leaders for more than seventy years.

Ecological experiments

I was blessed to have an opportunity to visit Hoover Dam at the border of Nevada and Arizona in the United States of America on May 6, 2024. I witnessed human ingenuity and efforts that use parts of creation. A dam was built to provide water for the inhabitants of desert-type regions. The arid lands with water scarcity can become the source of life for the locals and the natural habitat. The regions have also become tourist destinations and areas to take pride in by Americans.

Likewise, the preservation of the Grand Canyon in Arizona is not only a tourist destination but also one of the natural wonders and National Parks of the United States of America. As much as I can remember, a video presented the truths about the tribes who had lived there for many generations.

Connection with the lives and beliefs of native peoples

For the native peoples, there is an eternal connection between creation and their lives and families, as well as between their gods and themselves. According to their beliefs, their gods provide for their needs through creation. They believe that supernatural beings care for them and their creation and want them to live in harmony with the gifts given to them.

Despite rapid cultural and lifestyle changes happening rapidly in our postmodern world, the native tribal peoples around the Grand Canyon still preserve their cultures and lifestyles, providing necessary care and protection for creation.

Christians' specific manners of living can make a big difference in the world where constant changes in lifestyles have caused a lot of damage to the whole of creation. With progress in human development and changes in lifestyles come revolutionary changes in human life and God's creation. Consequently, these changes also impact people's understanding and practice of spirituality.

Evidently, since the Age of Enlightenment, people have discovered the power of reason and have used it for their lives. Reason is indeed human, Christian, and necessary because it is one of God's gifts to everyone. Nevertheless, some people use reasoning power in many unfortunate ways to fulfill their own inordinate and insatiable human desires. Sadly, many have yet to realize that the proper use of reason can help everyone find meaning and direction in life that are essentially rooted in our faith.

Wisdom from the life of St. Augustine

One of the best examples of a proper use of human reason is the conversion of Saint Augustine. Christians commonly know that the preaching and character of the intellectual giant St. Ambrose of Milan converted Saint Augustine. Before his conversion, St. Augustine searched for truth with the influence of Manichaeism (a religious or philosophical dualism). St. Ambrose's preaching helped the life of Augustine to turn his life around.

Saint Augustine with his mother, Saint Monica, by Ary Scheffer (1795–1858)

For Augustine, St. Ambrose's intellectual capacity was irresistible. After listening to St. Ambrose's sermons, "Augustine found himself becoming deeply interested in the content of what he heard. He concluded that Christianity could be intellectually respectable." (20) From his interest and fascination in the preaching of St. Ambrose, Augustine "witnessed a more astonishing conversion experience of Victorinus, [who was] one of the bands of scholarly defenders of paganism in Rome" as well as "the wonderful life and miracles of the hermit Anthony." (21)

We see here how St. Augustine's proper use of reason changed his life forever and consequently impacted and changed others' lives, too. It is fascinating to learn about the lives of St. Ambrose and St. Augustine, mainly how they correctly used their reasoning capabilities. Both proper and improper use of reason have corresponding effects.

We can inspire each other

How Christians care for others inspires many to become Christians and empowers them to care for others in the same way. For example, even many of my Buddhist relatives and friends have become Catholic after listening to my preaching and the stories I shared with them. Many have also become Catholics after encountering caring and loving Catholic priests and religious sisters who take risks to care for God's creation and work for the common good.

Christian love and care for others and creation sometimes go beyond human comprehension. To provide another example, some of my Buddhist relatives and friends even wonder how missionaries can survive in foreign lands where they have no families, relatives, or friends. My response to their questions and curiosities always surprises them because I always say to them that God appoints people in our lives to care for us wherever we are. In the Scriptures, God even appoints angels to guide and protect the people He appoints to work for His Kingdom. (22)

I fondly remember my encounter with a Myanmar national on a flight (Korean Air) when I was on my way back to Myanmar from the United States in 2016. He and I were seated next to each other on the plane en route to Myanmar. When he saw me well-dressed and with a crucifix around my neck, he asked, "Are you a Christian?" I told him that I was not only a Christian but also a Catholic priest. As we continued our conversation, we exchanged stories about our lives spent abroad (his life in South Korea and mine in the United States of America) and our planned vacation with our respective families in Myanmar. He joyfully remarked, "I love my work in Seoul, South Korea because I have witnessed and experienced how Christians in Seoul, care for overseas workers they don't even know. My Christian employers in South Korea have treated us justly, equally, and respectfully."

He emphasized that his employers in South Korea are different from his former employers in Myanmar, who happen to be non-Christians. It doesn't mean that Christians and Catholics are better than others. I want to highlight Christian duties and responsibilities embedded in their genes.

Finally, we parted at Yangon International Airport to meet our families, who were waiting for us with teary eyes. I told him, "It's great to meet and talk to you today. Enjoy your time with your family!" He likewise wished me well.

Throughout our conversation, I could see how he used his reason to understand the faith evident in Christians' actions. This one example of using human reason properly to understand human experiences gives a glimpse of great things that can happen under the care of Christians and other people of goodwill. How could human reason, which is very powerful, be properly used?

Faith and reason complete each other

Unless human reason is purified by faith, it can disrupt harmonious relationships between God and humans, between people themselves as well as between people and the rest of creation. It is very unfortunate to witness the fact that many people fail to see how complementary faith and reason are. When faith and reason are properly used, they complete each other, and life becomes more meaningful. Human endeavors become more practical, and people become more successful.

Nevertheless, many use pure faith or reason alone to live their lives. There could be many reasons behind such mindsets and attitudes. For example, the mindset and attitude of young people in our times are very fascinating. The environment they were brought up in must have had a significant influence on their worldview which is often different from those of older generations. There must be a significant difference in the way they think now, and the way older people did in the past.

Regardless of changes in everything in every generation, the power of faith and reason remains the same for Christians and people of goodwill. Faith and reason are needed for humans to live their lives to the fullest and to take care of God's creation. It is beneficial to reflect on the papal encyclicals on the relationship between faith and reason to clarify and understand the role of faith and reason in our lives.

Pope St. John Paul II, in his encyclical letter Fides et Ratio, explains the role of faith and reason in Christians' lives. Faith and reason are profoundly and indissolubly linked and united. Faith causes no interruption, let alone disruption, in freedom and range of action. Faith instead deepens our reason to see how God works in life's events.

The human mind can think through, discern, and decide properly and more effectively under the guidance and direction of God. Reason helps us to see the right path in life. Being enlightened by faith also strengthens us and allows us to persevere in achieving our goals. Separation of faith and reason weakens the human capacity to know ourselves, others, creation, and our Creator properly and effectively. (23)

Our strength lies in our differences, not only in our similarities

What could be the reasons behind the different phenomena of unique mentalities and attitudes of people of different generations? It is evident that every generation has unique ideologies with corresponding expressions and lifestyles. Young people's understanding of faith, spirituality, and their practice also differ from older generations.

I remember my homily on the sign people asked Jesus to give (Luke 11:29-32). I pointed out the changes happening in our lives, especially in the attitude and mentality of older and younger generations. After the mass, a woman friend of mine approached me and said with a great smile that things have changed from one generation to another. Your generation and mine are not the same.

We have many similarities as well as differences in every aspect of life. Reflecting on the spirituality of the gospel of Matthew, Hartin argues that the

"characteristic of young people in today's world is an honesty in acting according to their convictions. Inwardness captures the modern person's need to reflect upon his or her own selfhood, a reflection that is distanced from dependence upon others when making decisions." (24)

Including God in our significant decisions

We make decisions after deciding on something we desire to do or the kind of life we prefer. We decide only after considering the pros and cons of anything we plan to do. We even pray over our plan. So, we use both faith and reason if we want to make any significant and life-changing plans or decisions.

Pope John Paul II celebrates Mass near Mount Adamello in Italy's eastern Alps (July 1988); photo: Gregorini Demetrio

The importance of faith in God, whom our innate human desires seek, and the role of reason in human life are well explained by St. Pope John Paul II in the opening statement of his encyclical letter *Fides et Ratio (On the Relationship between Faith and Reason)*. Truth can be known and attained only with the proper use of faith and reason. They serve as "two wings" that enable us to kn ow and understand the truth about ourselves. They are also indispensable help and guidance to fully know, love, and serve God. (25)

I recall here a young man who had to bring his motorcycle to a repair shop after mistakenly putting water into the gas tank. He said he did this because he believed his faith would transform water into gas. His faith was a blind faith that was not guided by reason. Many of us may sometimes use either faith or reason alone in doing things and practicing our faith. We must have reasonable faith and a faith-filled reason to become the best version of ourselves and live our lives to the fullest.

Most of the time, we try to use reason in making daily choices and decisions about our daily activities. Indubitably, we are also enlightened by our faith in every good action. Our faith and reason enable us to do the best we can for God, others, and ourselves in and through others. Therefore, applying our faith in God in concrete daily circumstances and chores is essential to becoming credible witnesses of Jesus and the messengers of the Blessed Mother who appeared at La Salette on September 19, 1846.

Why is God always watching over us?

We may often ask, "Why does God repeatedly remind us of our Christian duties and responsibilities?" We may have dulled our reasoning ability slightly differently or even contrary to how it was initially given to us. I would like to believe that there are many contributing factors to the use or misuse of human reason, considering the rapid changes happening in every aspect of our lives and every sector of human society. Every generation poses different challenges for Christians and people of goodwill everywhere.

It is never easy to be a Christian in every age. According to Thomas Merton, one of the great mystics of the twentieth century, to be a Christian means to awaken to the real presence of God in our inner self. Christians must live a charitable and just life and acknowledge and support everything that brings unity among people. (26)

For Christians, the reality of God and the community God wants us to build are always present in Jesus Christ's life and ministry. He always speaks of the God who loves us and everything God helps us to accomplish in life until we attain eternal life. God also provides everything to help us do His will and sustain every aspect of life.

Creation is God's gift to all people

One of the essential things God provides for the life of His children is creation. Jesus often speaks of how creation works to generate, regenerate, and sustain itself and provide enough resources for all creatures. He assures us of God's unconditional and perpetual care for all of us. Perhaps many of us have forgotten that there is nothing to fear in this life because our heavenly Father takes care of us. (27)

The absolute assurance of God's care and protection is our strength to live our lives fearlessly. Living a life without fear is challenging but essential because only by living in freedom from fear can we truly change our lives for the better and eventually for the best. Yes, we all need to be free from all forms of fear so that we may be converted daily to the Lord.

Jesus said: "Do not fear... for your Father has given you the kingdom"

We all have experienced and know that fear does a lot of damage to our lives. For instance, everyone afraid of wanting the necessities of life can indulge in the accumulation of things. This eventually leads to greed that causes the imbalanced use of natural resources, even at the expense of others' needs. Even though God provides every natural resource for all in and through His beautiful creation, greedy people wouldn't necessarily care about the welfare of others but theirs alone.

The excessive accumulation of material things by a few people worldwide is very alarming. They limit the freedom of others by using their freedom excessively by their limitless accumulation of more things. They mistake authentic human freedom with a license that entitles them to do everything they want. Many, particularly greedy people, fail to see that "creaturely freedom is part of the given of creation, and this includes our responsibility for the future." (28)

What should we do with our freedom?

Imbalanced freedom and responsibilities in society and the family hamper healthy relationships. It is so enlightening to hear Pope Francis speak of a long-lasting effect of responsible freedom that calls for "intergenerational solidarity" (29) that will help us to possess an "adequate anthropology." (30) Only by understanding the needs of the human person, of the society they live in, and the influence of cultures can we thoroughly work on the development of every human need.

When we speak of human development, we must see and deepen our appreciation of it in the context of integral human development. We must understand that authentic human freedom is part and parcel of integral human development. God's eternal plan for all of us is that we enjoy living in the freedom He has given us. Only in that freedom, as Pope Francis reechoes the Judeo-Christian view of creation, we see creation as the reality that reveals the loving plan of God. That plan is none other than that in which "every creature has its own value and significance." (31)

Contrary to the general perception and belief of many in society, "the effects of human actions are not confined merely to the human world but have a ripple effect across nature." (32) Many still have to be educated about the impact human activities have on God's creation. Only their knowledge of the consequences of their actions may help them change their way of thinking, doing things, and living their lives.

What must we do if our actions affect the rest of creation? Ecumenical Patriarch I and St. Pope John Paul II made a joint statement on the necessity of a change of heart and, consequently, a change of lifestyle. They believed that "an inner change of heart [alone] leads to a transformation of lifestyles and patterns of production and consumption." (33) Two great and holy men understood and taught us that "a genuine conversion in Christ will enable us to change the way we think and act." (34)

We can do something very concrete with our converted and transformed heart to care for God's creation that is affected by human activities. In our time, a person's heart can be partially poisoned by material things. Our heart can also be overwhelmed by inordinate desires. Unfortunately, it often says yes to anything that it is accustomed to, like all of us who always say yes to our favorite music.

The Venerable Fulton J. Sheen, in his book The World's First Love: Mary, Mother of God, writes that "the heart has its secret melody, and one day when the score is played, the heart answers: 'This is it.'" (35)

The Venerable Fulton J. Sheen, Roman Catholic Bishop and early television preacher during the 1950s and 1960s; photo: Fred Palumbo

When will the human heart understand its logic of beating – beating for its Creator? Every human person must realize both their creator and their creatureliness. According to Rudolf Bultmann, Pauline theology explains that God's concern for humans is that every human person lives and the Earth provides for their needs. The sole reason and purpose of creating the whole world and everything in it is to allow humans to properly use God's creation for the well-being and benefit of every human person. (36)

The kind of love God has for us is the love that cares and provides without any reservations for Himself. What happens when humans accumulate everything that creation offers, although they do not need it? We shall explore the consequences of reckless human actions born out of insatiable human desires.

Unjust systems anywhere threaten justice everywhere

To address human needs, we need to evaluate and change unjust systems that deprive fundamental rights and needs of humans and the rest of creation. Individual conversion sometimes doesn't suffice to overhaul everything that benefits only a few but causes the suffering of many. Therefore, structural reform is often indispensable if our dreams, plans, and actions are to impact a long-term change for the better. Alongside continuing changes in society, the theological underpinning of the reality of change in every aspect of life has developed.

"The liberation theology of the later twentieth century, building in part of [the] personalist theology of grace and conversion, calls attention to and demands the reform or dismantling of the social structures that undermine community and obstruct communion." (37)

What can we do to make changes for the better happen? We must rely on God's grace and cooperate with it. Believers in the Triune God and the people of goodwill believe that only God has the power to cleanse us of our sins and to help us to convert to Him as He reminds us in the book of the prophet Isaiah: "Come now, let us set things right, says the Lord: Though your sins be like scarlet, they may become white as snow; Though they are red like crimson, they may become white as wool. If you are willing and obey, you shall eat the good things of the land." (38)

What are some of our questions about conversion?

Many of us may have many questions when it comes to conversion. We may ask ourselves and others these questions. What must we be converted from? How often do we need to convert? To convert is part and parcel of human growth, especially in a mature Christian life. Mature people and Christians tell us that they underwent a long and sometimes painful process of conversion. The Church teaches us the conversion process and its effects on personal and communal lives. We hear the teaching of the Church on the essence of conversion and its process.

St. Pope John Paul II, in his apostolic exhortation, Familiaris Consortio (on the role of the Christian family in the modern world), explained well how genuine conversion happens in the lives of Christians. He encourages us that we need to undergo conversion all the time so as to detach ourselves from all evil desires and things interiorly. On the other hand, we need to align our lives to everything that is good to make concrete and practical steps to move on.

Our personal and social lives must be gradually developed by integrating God's gifts and absolute and unconditional love. Christ's love that grows and matures in every person and the whole of the universe must lead every fiber of the family and of society into the holistic comprehension of the essence of God's love made manifest and tangible in Christ Jesus. (39)

Our personal conversion can affect many others

Conversion for Christians is not reserved for a private life. It is not about personal holiness. It is inclusive and concerns the community to which they belong. Therefore, holiness concerns everyone and even the rest of creation. We are holy or unholy only in relation to the rest of humanity and the whole of creation.

For these reasons, when we speak of conversion, we must remember that Pope Francis urges us that "Christian conversion demands reviewing especially those areas and aspects of life-related to the social order and the pursuit of the common good." (40) Most of us in this generation are often affected by the lures, tricks, and traps of humanism, materialism, consumerism, and relativism. Many people find it challenging to realize the realities contributing to many social ills. Very few dare to acknowledge and overcome them. Some know what is happening in their lives but have neither the resources nor the capacity and attitude to avoid, resist, and overcome them.

Our life is our pilgrimage back to the Father

My visit to the Holy Land (Israel and Palestine) in 2022 was very touching for me and those 25 parishioners of St. Ann's Church in Marietta, Georgia, and friends who joined the pilgrimage. I observed that many of us, like most pilgrims, go to the Holy Land where Jesus was born, grew up, preached the Good News, healed the sick, drove out the demons and raised the dead, suffered, died, and rose again from the dead, with great faith to experience the presence of God in concrete historical settings and spots. Many also go there with great hope for the conversion of their lives and those of their loved ones.

I found visiting, praying, and celebrating Masses on the holy sites very touching. I could see the interplay of relationships that led to Jesus's suffering, death, and resurrection. Although Jesus has won the victory over sins and death, the structures of sin in every generation continue to impact, challenge, and attack us, God's beloved children.

Therefore, there is always a need for conversion. It is good to remember that God's promise to King Solomon is His promise to all who undergo conversion. God said to Solomon,

"If then my people, upon whom my name has been pronounced, humble themselves and pray, and seek my face and turn from their evil ways, I will hear them from heaven and pardon their sins and heal their land." (41)

Reflection Questions:

1. How do I construe what conversion is according to the truths of Jesus' teachings?
2. How often do I use God's Word to help me undergo stages of conversion daily?
3. How willingly do I try my best to undergo the process of my conversion to the Lord?
4. What do I need to be converted from?
5. How do I perceive socio-political and relational conversions?

Chapter Three

Welcoming Change in Our Attitudes

"God looked at everything he had made,
and found it very good."
(Genesis 1:31)

Lord, I lift my heartfelt thanks to you
for the gift of creation.
Make me an instrument of Your loving presence
in every environment.
May both human and natural environments
be safe, healthy, and beneficial
to all who live in them.
May your glory and power be visible
in our daily relationships
with everything you have given us.

Original Beauty and Human Ingenuity

Our human environments either protect and preserve or destroy and devastate natural environments. My experiences have taught me much about the link between the two that we cannot neglect or deny.

Since my arrival at St. Ann's Catholic Church in Marietta, Georgia, in the United States of America, on September 27, 2014, I have seen and visited many beautiful places in the United States. I have experienced and realized how captivating they are. They have imprinted lifetime memories upon my mind and heart. I will never tire of cherishing and sharing beautiful memories of my immense experience of God's beautiful creation.

Holy Ghost Catholic Church, Maui, Kula, Hawaii, the "Octagon Church"

One of the best places in the United States that people would often like to visit is Hawaii. They even fondly call it "a paradise." Honolulu has many historic sites, such as Pearl Harbor and the Polynesian Center, while the cities of Maui and Lahaina are more intact in their natural beauty. After my 47th birthday, I visited Maui, Hawaii, the United States of America, on May 23, 2023. As I enjoyed the beauty of the island, I was blessed to participate in the preparation of the priest and his parishioners for the feast of the Pentecost. It was the feast of the parish that had a lot of Portuguese culture and traditions.

One of our Myanmar La Salette Missionaries who joined the Philippine Province of Mary, Mother of Hope, together with his beloved parishioners of the Holy Ghost Catholic Church, was preparing for the Feast of the Pentecost. The church of the Holy Ghost is built upon the northwest slopes of Mount Haleakala. Since the 1890s, the church has been an Upcountry Maui landmark.

The Octagon Church (the Wedding Cake Church) amazed me with its original architecture and beauty that were popular in the 1800s.

A Portuguese Feast in Hawaii

After showing me around and inside the church, they introduced me to the making of Lau Lau, which is a Hawaiian dish (originally it was prepared to feed the poor) made of pork, beef, and (butterfish, if available) wrapped with Lau Lau leaves and Ti leaves. Lau Lau has to be steamed and eaten with rice. They usually feed two or three thousand people on the Day of Pentecost. They have celebrated this feast for more than a hundred years.

Drawing from the Portuguese Tradition that began during the time of Queen Isabel (Saint Elizabeth, 1271-1336), the Portuguese who came to Hawaii in the late 1800s continued to practice their devotion to the Holy Ghost (Spirit) and celebrate the traditions of St. Elizabeth. Until today, their descendants and Catholic faithful feed Lau Lau to all who come to the feast.

Protecting Hawaii's natural habitat, vegetation, and species

As I helped them make Lau Lau using native vegetables and leaves, I was reminded of filling out the form on the plane before we landed at the Maui airport. The questions were about bringing seeds, fruits, and plants into Hawaii. I am sure the local government and the locals want to protect the land from any harm insects and pests could cause to the natural habitat and vegetation. They try to protect the environment, crops, and species at all costs.

I found that every food and drink they offered was very palatable and refreshing. The food on the island is indeed fresh, tasty, and healthy. My experience there deepened in me the core value of environmental protection. I am sure that we all will have fresh and healthy food if everyone shares their responsibilities and takes part in protecting and taking care of creation everywhere all the time.

I also had an opportunity to swim at the bank of the North Pacific Ocean in Haiku-Pauwela, Maui, Hawaii, in 2023. Using a diving mask and a snorkel tube I could see the beauty of underwater creatures – including a variety of fish and giant turtles. A trumpet fish amused me; giant turtles were fun to swim with. I also found a small sea urchin and released it back into the ocean after holding it with the strap of my diving mask.

Gerard Manly Hopkins wrote: "The world is charged with the grandeur of God."

The natural environment in Maui just amazed me. Lying down on the ocean bank while being splashed by the waves and listening to the sounds of the waves moved me to be thankful to our loving Creator God more than ever. What would happen if we lost the natural habitat of marine creatures? I surmise that the damage and the eventual loss of both the beauty of the aquatic world and its inhabitants would endanger our own identity and livelihood and, eventually, the human species.

With the help of my priest friend, we went up Mount Haleakalā, which means the house of the Sun in Hawaiian, on May 29, 2023. The summit is 10,023 feet (or 3,055 meters) above sea level. As we drove up the mountain on the Island of Maui, I told my friend that the flowers in full bloom along the roadsides looked like those that were intentionally and masterfully planted by the best landscapers. They were just gorgeous. To my amazement, he said these beautiful flowers would be gone in a few weeks because of the summer heat.

Mount Haleakalā at over 10,000 feet overlooks a giant crater on the Hawaiian Island of Maui; photo: Navin75

He said that I visited Maui at the best time of the year. Contrary to the human tendencies, urging me to take home memorabilia, my friend warned me that neither a piece of rock nor a plant must be taken away from there. After a few minutes of strolling, so true to his warnings, I found all the signs that said exactly what my friend told me. The only thing I was allowed to do was to take pictures and videos of everything that caught my attention, especially the crater, the flowers, and the birds. I even saw a beautiful bird resembling a quail on Mount Haleakala. The mountain has the freshest air and the cleanest environment I have ever experienced.

My stay in Maui taught me invaluable lessons about human identity and responsibility. When humans know and find their identity and attend to their duties, the natural environment or creation also has its own identity and helps people fulfill their purpose in life. All the creatures of God live and have been together in the love of God that gives life to all and leads all to eternal life. We, animate and inanimate alike, belong to the same earthly home. Pope Francis coined the term "our common home" to inspire, invite, and encourage people to care for God's creation lest all creatures become extinct.

Appreciation for our natural environment

Many of us may not be inclined to take care of creation but love to enjoy its beauty. I was blessed to visit San Diego, California, U.S.A. in 2024. As I enjoyed a bus tour around the zoo, I read the instructions concerning how to enjoy but avoid harming the natural habitat or animals. Similarly, I also saw signs prohibiting visitors from getting too close to the animals. While driving the tour bus, the tour guide explained how habitats from around the world are preserved, and the animals are taken care of. From the smallest birds, such as male Bee Hummingbirds, to the largest mammals, such as giraffes, rhinos, and elephants who live in the zoo. It was my first time to see a polar bear, a cheetah, an Asian tiger, and a hippo. They all have the food and climate they need.

In 2024, I visited Big Bear Lake, California. A Myanmar La Salette missionary, Fr. Philip Makha Naw Aung, M.S., was driving as I enjoyed the scenic views of the San Bernardino Mountains. Once we reached the city of Big Bear, we visited the zoo. Although the temperature was ninety degrees Fahrenheit, I enjoyed seeing a black bear, a grizzly bear, a chayote, a leopard, and some eagles.

We need to take better care of God's creation. Our tour guide at the San Diego Zoo said that all the visitor's fees are destined to go to the care and conservation of the habitat and animals in the zoo. My visits to such places always inspire me to care for and love creation.

We only have one Earth

One of my favorite spiritual writers of twenty-first century, Ronald Rolheiser, O.M.I., in his book Wrestling with God, reminds us that the world we live in is "a place for all of us – humans, animals, insects, plants, water, rocks, and soil – to enjoy a home together." (42) As the Sacred Scriptures and the Catholic Social Teachings continually remind us, we all have inalienable rights. Fr. Rolheiser reechoes that truth:

"Physical nature too has inherent rights – namely, rights that are intrinsic to itself and not simply given to it because of its relationship to humanity. The earth is not just a stage for human beings to play on but is a creature of God with its own rights that humans may not violate." (43)

The change in human attitudes is needed now, more than ever, especially in our relation to the whole of creation. Our human indifference towards creation is never good news and is getting worse over time. Many fail to see and understand that life is all about relationships between God and His creatures. Human and salvation history tell us how crucial these relationships are.

Kathryn James Hermes, a religious sister of the Congregation of the Daughters of St. Paul, explains it well, "At its deepest level, history – whether that of a nation or that of a person – is about the relationship between God and us." (44) If we read it carefully, Holy Scripture has much to teach us about human relationships, let alone our relationship with God and His creation.

God's creative hand

After creating the Earth, water, vegetation, the sun and the moon, the stars and all creatures, God created humans and blessed them. He also entrusted everything to our care so that every good thing He created would be for His glory and the welfare of His people. (45) We see how good God's original design and plan for His created beings was.

Despite the irresponsibility of many, God continues to guide His creation through His creatures, especially through the cooperation of His faithful people and collaboration among the people themselves. St. John Paul II and Ecumenical Patriarch Bartholomew I, in their joint statement on the ecological crisis the whole world is facing, expressed their:

> "concern about the negative consequences for humanity and for all creation resulting from the degradation of some basic natural resources such as water, air, and land, brought about by an economic and technological progress which does not recognize and take into account its limits." (46)

On the scale of one to ten, what is our love for God and His creation? And how deep is our concern for the health and growth of creation?

The numbness of the human heart

Human indifference to nature's need and cry can lead to inhumane treatment of mother nature. Mismanagement of natural resources can lead to the mistreatment of nature that eventually causes the destruction of nature itself. To my understanding, it must be being overwhelmed by the struggles and sufferings of the time that makes many ignore or remain indifferent towards environmental degradation. When we are numb to the urgent needs of others — let alone to those of creation — we contribute to the ongoing sufferings of others.

Sadly, those who live in underdeveloped or developing nations suffer more from the impact of environmental crises than those who live in developed countries. They have no other options but to endure never-ending suffering caused by human and environmental illnesses. The people who live in many countries that rely mainly on agriculture suffer the most. Growing crops in arid lands can produce very little or nothing at all.

These two volunteers are participating in "The Great Global Cleanup," the worldwide effort to improve our Earthly environment during Earth Day

The tragic cycle of environmental degradation

I recall my meeting with the farmers in Myanmar who had been affected by environmental crises for many years. Some shared with me how landlords would exploit their rights and privileges. In my mission parish in Myanmar (2010-2013), one farmer said that Chinese businessmen and businesswomen (from mainland China) would use their lands for two years to grow crops and have the best produce. Sadly, the farmers would have inferior produce or none after the businessmen and businesswomen returned their leased lands to them. he result for the farmers was that they lived in deep debt and abject poverty due to the Chinese businessmen who used excessive amounts of pesticides and fertilizers that would exhaust the soil. Furthermore, the soil became so arid that even grass could not grow.

Even the animals—water buffalos, cows, and oxen they use for plowing the fields—are affected by the pollution of the air and water and the eventual destruction of vegetation. The animals have insufficient or low-quality food to the extent that their families' health is affected.

Their poor health is detrimental to human health in the long run, as both meat and milk are of poor quality for nourishment. Poor quality produce provides farmers with only a meager income, forcing them to obtain loans with a high interest rate either from the government or the wealthy of the area. This devastating chain of environmental degradation affects the whole of creation and creatures. Every aspect of the lives of the poor is always in great peril due to environmental degradation.

Efforts at environmental protection

In my first-hand visit to the Holy Land – Israel and Palestine –in 2022, it was astonishing to see the people in the Middle East building irrigation systems to benefit vegetation and crops. No fresh water was wasted –from showers and water used for washing clothes and dishes – but it was processed and reused for plantations. Our tour guide explained that wrapping banana bunches with plastic protected them from insects, pests, and birds. Furthermore, he explained that banana bunches must be wrapped because they do not use pesticides that drive away or kill insects or pests. This unique experience reminds me of the agricultural lands I have lived in or visited.

Taking extra care when using pesticides

Farmers use pesticides and fertilizers in many parts of the world, especially in poor nations that rely on agriculture. Experts must monitor safety measures for the health of creation and humans. In developed nations, local and national governments protect the natural and human environments.

A businessman in Myanmar explained to a customer the difference between his two plantations. He said the plantation with ugly and small vegetation and fruits was for his family. The produce was small and not pleasing to the eyes because he never used fertilizer or pesticides. On the other hand, those good-looking plants, fruits, and produce were for sale. They were cultivated as such to attract and impress customers.

The dark side of this reality is scary and detrimental to our health because too many pesticides and fertilizers can eventually destroy the soil, the air, and the water around the plantations, and, of course, human health. This is one of the major causes of extreme and continuing poverty in countries that rely on agriculture.

Farmers in developed nations use pesticides and fertilizers with extra care so as not to destroy the ecosystem. In the United States, farmers grow fruit, vegetables, fish, and farm animals with the proper use of fertilizers and pesticides so that neither the health of their crops, animals, nor the health of consumers are affected. I remember a friend sharing his experience of caring for his lake and the fish in it. Pesticides were used to get rid of weeds that affect the lake's water. However, the treatment will take a week or so to normalize the water so he can fish and eat his catch once again.

The environmental crisis and human migration

The effects of God's eternal and unconditional love are evident in His creation. According to the prophet Isaiah, God had a perfect design for the creation of humans. His creation of the heavenly bodies and the earthly beings had a quintessential beauty. Humans were given all living things to enjoy and care for (45:18). No masterpiece of art will ever be comparable to God's prototypical blueprint of creation.

o remedy the brokenness of human hearts, God, through the prophet Ezekiel, reminds us of God's readiness and willingness to renew our minds and hearts despite our stubbornness. He could remove anything that can harden our natural hearts that He beautifully and perfectly designed. God always tried to rekindle the true spirit in all of us so that we might learn to love and care for one another and the whole creation as well (36:26).

It is our human and Christian duty to educate ourselves on environmental degradation, which is one of the leading causes of human migration.

"The International Organization for Migration estimates that more than 214 million people are migrating around the world today. This means that one out of every 33 people worldwide lives away from their homeland. Approximately 42 million migrants are forcibly uprooted, including 16 million refugees outside their countries and 26 million who are internally displaced." (47)

We have many Myanmar refugees in many countries around the world, as well as thousands of internally displaced people in Myanmar. They are in such situations due to conflicts that arise from the greed of the powerful and the rich, who exploit the lands of simple farmers and cause conflicts. I remember many poor farmers losing their farms to the landlords when they could not pay the debt due to the poor produce of the lands and the high interest rate for the loan. These economic conditions caused by the ecological crisis and the monopoly of the powerful and the rich over the natural resources of Myanmar turned many farmers into coolies.

Even some of my relatives lost their farms and properties to the dam built by the military and its cronies. Farmers like my relatives never received any compensation for the loss of their lands. These conditions open the gate to unrest and armed conflicts, too. o flee poverty and conflicts caused by both human and environmental crises and to build a better future for their families, people migrate into other countries even risking their identity, dignity, and lives.

Immigration caused by environmental problems

Migration has lasting effects on the lives of refugees and migrants. For Catholics, it is more challenging because migration can fight against the core of our faith life.

We Catholics believe in the One God, profess the one faith, and practice the teachings of Jesus. St. Paul teaches us that truth in his letter to the Ephesians. (48) We who believe in One God, embrace One Faith, receive One Baptism, and follow One Lord are convinced of only one human race. Being one with the One God and the One human race, we are also challenged by the question of being One with the rest of creation, too.

We all must have realized that environmental problems cause damage to the life of every creature and the whole of God's creation. The ailing environment forces many to flee their homelands for safety, livelihood, and a better life for their families in foreign lands. Endangering God's creation is endangering our own human family, faith, and the church. The faith that unites us is the only thing refugees and migrants can hold on to.

Sometimes, even God seems absent for many in the more challenging situations in life. In her reaction to the military coup in Myanmar in 2021, a friend of mine said, "God is not in Myanmar anymore. He has gone abroad to find a better place to live!" Her reactions have realistic points because political chaos and environmental damages do have lasting effects on the lives of many, especially on the poor and rejects of society. The poor become the victims of consumerism, materialism, and hedonism. Consequently, those suffering from the effects of poverty are always at the mercy of the powerful and the rich. God, faith, spirituality, and the Church are the only recourse for the victims of indifference. How shall we address human indifference in our times that contributes to human and environmental illnesses?

A local church as a refuge for migrants

Church environment is indispensable to help protect, preserve, and promote human values facing existential threats. Evidently, the lives, livelihoods, faith, and cultures are at their utmost test when migration happens. Here is my experience of working with Myanmar refugees and migrants in the United States. It will enlighten the minds and hearts of many as they become aware of the human indifference that affects every aspect of life. The lives of many people are changed forever due to human indifference.

Stephen O'Kane, Staff Writer of Georgia Bulletin, wrote an article, "Burmese Catholics Find Support in Parish, Deacon," about the Burmese refugees who found a home in the Archdiocese of Atlanta, Georgia, on November 26, 2009:

"The Catholic Burmese population throughout North Georgia is rising at a rapid rate, and Deacon Peter Swan of St. Philip Benizi Church, Jonesboro, has been there every step of the way to minister to the group." (49)

Since the beginning of their settlement in Georgia, the Myanmar Catholic Community of Atlanta (M.C.C.A.) has been without any Burmese-speaking priest assigned to them for their pastoral care. After more than fifteen years, the faithful and loving God in His time fulfilled His plans for His beloved people of Myanmar, who have found a home in the Archdiocese of Atlanta. Even amid sad events in human history, such as the pandemic and the political turmoil in Myanmar that changed all aspects of life, God still accompanies His people in various ways. God appoints good people to accompany His beloved children, especially the refugees and the forgotten of human society.

At the outset of the pandemic, I had been *administering* celebrating sacraments with the Myanmar Catholic Community of Atlanta every other month and as needed without any interference with my La Salette ministry at St. Ann Parish, Marietta, Georgia. Fr. Raymond Cadran, M.S., was so good to let me journey with my fellow Myanmar Catholics for almost two years. Better things came into the lives of M.C.C.A. members after many informal meetings with the past and present leaders of M.C.C.A. Their dream to have a Burmese-speaking priest came true when Archbishop Gregory John Hartmayer, O.F.M., Conv., of the Archdiocese of Atlanta, Georgia, requested that I become the chaplain for the M.C.C.A. in September of 2022. When the M.C.C.A. heard the good news and saw the glimpses of the official presence of a Burmese-speaking priest in their midst, everyone was overjoyed and thrilled, and some even became very emotional.

Greater things were already in store for them, without their awareness – let alone their knowledge — because Archbishop Gregory Hartmayer, O.F.M. Conv. and Deacon Peter Swan (the only Burmese native) of the nearby St. Philip Benizi Church, the Archdiocese of Atlanta, knew each other for a long time before the archbishop became the bishop of the Diocese of Savannah, Georgia.

A long-standing friendship

I concur that the friendship between the archbishop and the deacon has continued and been deepened for over many years. Their friendship facilitated every canonical step, making it easier and a great help in taking further steps to finalize the appointment of a chaplain for M.C.C.A. Although I have been helping the M.C.C.A. occasionally, that is, Christmas and Easter Masses, my ministries ~~to~~ them were very limited. We can be assured that our God is indeed a God of surprises.

The birth of new hope for Burmese Catholics in Atlanta

God uses every human event to reveal His glory and carry out His salvific plan for all. For Myanmar refugees and immigrants, God answered their prayers on March 1, 2022, when their more-than-a-decade-long request for a Burmese-speaking chaplain was finally granted.

When the leaders and the community requested me to be their chaplain, I assured them of my availability if the La Salette Community would allow me to take it up. With the help of God's grace, the blessing of the Province of Mary, Mother of Americas and the appointment made by the Archbishop Gregory Hartmayer, I officially began my chaplaincy to M.C.C.A. on March 1, 2022.

Burmese Catholics gathered for their National Conference in Atlanta recently. A Buffalo, N.Y. youth group sang a Burmese gospel song; photo: Michael Alexander

The community formed the Myanmar Pastoral Council (M.P.C.) to help me, their chaplain, to understand, and respond to the needs of M.C.C.A. members. All ethnic groups and tribes are represented in the M.P.C.

To ensure that La Salette spirituality is explained, and practiced, they pray the Memorare to Our Lady of La Salette and the La Salette Invocation at the end of every meeting. They continue to discern and learn about the pastoral needs of their membership.

Since the foundation of M.C.C.A., the primary pastoral need has been the celebration of Sacraments in Burmese, the official language of Myanmar, in two parishes where two ethnic groups belong. The two parishes, Corpus Christi Catholic Church in Stone Mountain and Holy Cross Catholic Church in Decatur, have been graciously helping Myanmar refugees and immigrants for more than a decade. I am very grateful to them for the welcoming and accommodating spirit they have shown to countless Myanmar refugees and immigrants. We see vividly here how relationships make a big difference in the lives of many.

A new ministry benefiting our Burmese Catholics

Since day one, I have been involved in engaging new ministries. As people learned about my official appointment for their pastoral needs, more exciting things are happening, such as invitations to their birthdays, wedding anniversaries, and graduations. I also taught a Pre-Cana Class to a young couple who got married recently.

Most refugees can benefit greatly from orientation sessions

I together with 50 people (ethnic leaders and youth) also visited the refugee camp called Jubilee where refugees are given orientation and inculturation for six months when they first arrive in Georgia from any country around the world. It's very exciting to see the welcoming sign that has greetings written in many different languages of the world. They, likewise, prayed at the cemetery where the diseased refugees were buried. Everyone has a role to play in every human community.

M.C.C.A. had an official welcoming ceremony for me, their chaplain. About 500 people of all ethnic groups came to celebrate the reality of their dreams. Since the com-munity is composed of young families, we have more young people. There is no comparison between their past lives in Myanmar and their new lives here in the United States.

Education is a sign of hope for the children of migrants

Many families celebrated their children's graduation from college and advanced degrees for their children for the first time in their families family. This was very emotional for them because their relatives and friends back in Myanmar could never have such opportunities for education and academic accomplishments.

One family was overjoyed upon their daughter's Valedictorian Award. Ja Seng was awarded with a four-year scholarship to High School. I foresee many more great events and future blessings coming from M.C.C.A.

Community is also central to immigrants

What is needed is having the right relationships with others, so much so that we help one another to see beyond the bend and expand the horizon of our lives. In June 2022, our new immigrants attended the Archdiocesan Eucharistic Congress.

It was the first time to provide a Burmese Track on the Eucharist. I was one of the presenters for the Burmese Track. My topic was: The Eucharist, the Only Remedy for this Wounded World. It was based on "Come to Me," the theme of the 2022 Eucharistic Congress.

Burmese Catholics long for the Eucharist and want to learn more about its meaning and richness. Their longing for the Eucharist became more ardent due to the pandemic. During 2022, some of our members attended the National Conference of Burmese American Catholics (N.C.B.A.C) in the diocese of Owensboro, Kentucky. It featured a Priestly Ordination of a Burmese-American deacon. He is the second Burmese American deacon to be ordained for the American church.

When relationships are right, better things come into our lives. M.C.C.A. also found many reasons to celebrate many great and historical events.

- October, 2022: 15-Year Anniversary of the Myanmar Catholic Community of Atlanta; Archbishop Gregory John Hartmayer, O.F.M., Conv., presided and was inspired by the congregation who sang well together with the M.C.C.A choir.
- The members of the M.C.C.A were very thankful for the Archbishop's fatherly support to their every effort ~~to~~ bring about unity among the seven ethnic groups from the dioceses of Myanmar.
- The gathered community thanked the Province of Mary, Mother of Americas, for allowing me, a Burmese-speaking La Salette priest to shepherd the Myanmar Catholic Community of Atlanta in the Archdiocese of Atlanta. Since I was appointed their chaplain, the community has become more united.

No more war!

On May 3, 2024, I, together with the Burmese American Catholics for Justice and Peace, was privileged to participate in a meeting with the Peace and Justice Committee of the United States Conference of Catholic Bishops to discuss how to help the local church of Myanmar. Myanmar citizens still have high hopes for ending many decades of authoritarian rule.

Myanmar Junta Air Force bombs a church in Kayah State, May 2024; photo: MPATV

Conversely, the hopes and dreams of Myanmar citizens were shattered on February 1, 2021, when the military junta staged a coup d'état. Many lives and livelihoods have been sacrificed in the efforts to redeem the country from the military regime. It is now more than three years that innocent people from all walks of life have been living in fear, anguish, uncertainties, and scarcity of basic needs.

The lingering devastation from Myanmar's Civil War

Innocent citizens are in dire need of food, shelter, and medical supplies. Furthermore, the conscription of citizens aged 18-35 had devasting effects on the lives of families who had been struggling to survive during the Civil War. The local church has also been scrambling to help meet the needs of the victims of this war, particularly the internally displaced people.

To provide a vivid example of devastation, the smallest but predominantly Catholic state of Myanmar, Kayah, has suffered the most from the military assaults. Many villages and towns in it have been leveled, and people have been internally displaced more than three years now. On November 27, 2023, the bishop of Loikaw diocese (Kayah state), the clergy, and the religious sisters, together with the orphans, infirm, and disabled people, had to flee the diocese after the military took over the cathedral and all religious facilities.

The bishop, priests, religious sisters, and the infirm and disabled people have found refuge in the territory of the Karenni Defense Forces. Until August of 2024, the local people in Myanmar and the relatives of the immigrants in Atlanta, Georgia, said that most cities, towns and villages in Kayah state are no longer inhabitable due to the destruction of the infrastructures, livestock, continuing onslaught by the regime, and the ongoing battle between the military forces and People's Defense Forces. he number of people in need of humanitari-an assistance keeps rising as the depleting military becomes more desperate to safeguard its existence in Myanmar.

The gift of foreign aid from non-profits

The only glimmering hope for innocent civilians at present and in the post-revolution era is the foreign aid that comes from nonprofit organizations, one of which has been the Burmese American Catholics for Justice and Peace (B.A.C.P.J.). As far as we, the BACPJ, are cognizant, there are three in dispensable things. hese will help the people living in fear and anguish as the regime's strategic crackdown on those who resist their rule continues.

In collaboration with local Catholic churches in Myanmar, Catholic churches in the United States of America, and charitable organizations, we can do three essential things to help the post-civil war Myanmar to recover from the trauma of violence and rebuild their lives in freedom, justice, peace, and prosperity.

1. Shelter: No one is safe, and there is no safe place in Myanmar. Even my best friend, a La Salette Missionary priest who returned home to Myanmar from France, was arrested at a checkpoint. He was on his way to visit his family who had been hiding in the woods for more than two years. His release is uncertain despite the legal settlements the community has made with the regime.

   Many more arrests are still being made. And the most recent cruelty the military regime has imposed on the people is more sickening. Young people aged 18-35 are being conscripted into the military. Since atrocities and heinous crimes are still being committed by the military regime, people still need outside help.

2. Basic Needs: In 2023, with very little produce of rice and crops and no produce at all in many regions, due to the ongoing civil war, farmers remain wary of food scarcity in the following years. According to my family, relatives, and friends, the country has had a shortage of basic needs and medical supplies. Many will run out of money to pay the rent in cities, send children to school, buy food, and meet other basic needs.

3. Long-Term Support Programs: Many priests, religious men, and women have witnessed people's need for rehabilitation from traumatic experiences. Rehabilitation programs will be needed even after the country is free from authoritarian rule in the near future. The help of the BACJP and the social

and charitable arms of the universal and local churches will be indispensable to support post-revolution programs for the survivors to rebuild their lives, renew their livelihoods, and rebuild infrastructures such as schools, hospitals, and churches that the civil war has destroyed.

Building a new life

We see here that life is beautiful when we journey together, setting aside our differences but celebrating our uniqueness and collaborating, utilizing our God-given charisms, gifts, talents, and expertise for the good of all and God's glory. That is how we build God's family and His Kingdom. That's also the way of life of the La Salette Missionaries entrusted with the ministries of forgiveness and reconciliation for those on the margins of society and the Church.

As the plight of refugees, immigrants, and internally displaced people are affected chiefly by moral, political, and spiritual paralysis across the globe, we must find ways and means to address this life-changing issue in our time as the refugee crisis continues to rise.

From a spiritual standpoint, we need an integral spirituality that includes ecological spirituality. Such a spirituality can be lived out daily to combat global crises that affect the common good. To address these environmental issues that are some of the root causes of conflicts and unrests around the world, we must be aware of the sad reality of ecological crises caused by reckless human activities.

Our need for both economic growth and ecological protection

Jane Blewett, in her article "Social Justice and Creation Spirituality," enlightens us,

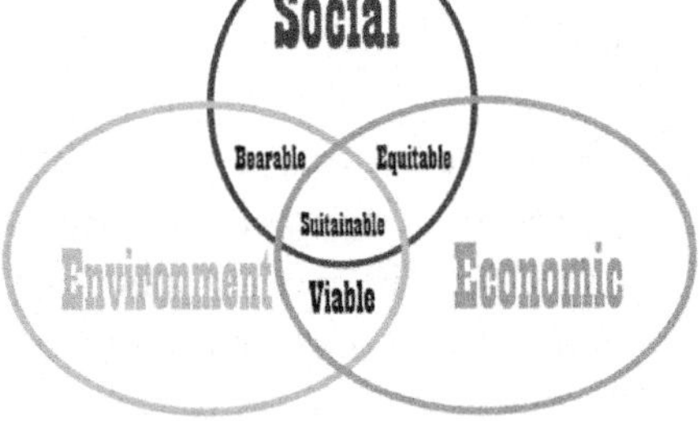

The diagram explains the relationship between the three theoretical issues

"At present, humans are like spoiled children striking blindly at all that lies in our path, as we destroy the rainforests, reroute rivers, level mountains, hardly giving a thought to the eons of generative life to which each gives testimony, nor the communities of living creatures that each sustains. Surely, the rainforests of Brazil with their approximately 50,000 species of trees (as compared to with about 700 in all of America), deserve more regard than to be chain-sawed into extinction. When we annihilate a species, we lose forever an expression of God, an experience of God." (50)

There must be a good balance between economic growth and environmental protection because both are either beneficial or detrimental to humans. In the long run, ecological health affects every aspect of human life, beginning with the air we breathe, the water we use and drink, the food we consume, and any other things we use.

I remember my grandmother allowing me to drink water in palm of her hands. She would get it from a stream as we went back home to our village after visiting her in a different village every year. In those days, water from any stream, creek, river, or well was clean and safe to drink, let alone suitable for plantations and crops.

Mother Nature has her own way of coping with our necessities

Furthermore, my parents and grandparents used to tell me to come home before four in the afternoon during the rainy seasons that lasted for four months. I recall how clouds formed in the sky and darkened at around four o'clock. As they always reminded me, it rained a few minutes after the clouds formed and the wind blew. Although not fully understood, I could see that God's creation has its way of taking care of itself to generate and produce food for all creatures. Once the rain stopped, people would go out and hunt for birds and animals who also came out to enjoy the cool weather at night. What a lovely life to live in Myanmar those many years ago!

In this age of ours, or what we call "a very developed age," rain is scarce, and water and air are polluted. Consequently, farmers have very little produce, and even when they have a lot to harvest, their produce is always poor in quality. We all can agree that ecological devastation is real but not yet necessarily extreme.

The Earth herself shows us how to care for creation

Although creation is already damaged, it isn't beyond repair. We can still help creation to regenerate itself by caring for it. Perhaps, we can imitate other creatures of God. For example, I was blessed to see chipmunks, squirrels, opossums, raccoons, rabbits, a herd of deer, snakes, birds, and insects roaming around, playing, or looking for food and savoring it in our backyard in Marietta, Georgia. They seemed to be spreading the message of our responsibility to care for God's creation. Those creatures give me a very strong message: "It is the Earth [that] gives direction to life forces, [that] nurtures, sustains, waters, warms, and cools." (51) All creatures of God benefit from the earth's produce.

Many of us might have encountered other creatures of God giving thanks to God in their own ways every day. Denis Edwards, in his article "Earth as God's Creation": The Theology of the Natural World, in Pope Francis' Laudato Si', beautifully states that "every creature has its word to speak. None is superfluous. Each can become a caress of God for us." (52)

God indeed does a tremendous job in creating the world because it provides everything we need – to praise, love, and serve Him in and through other fellow creatures and created realities. Pope Francis reflects even more profoundly on the richness of God's creation by re-echoing St. John Paul II's catechesis (January 30, 2002, 6), "God has written a precious book, whose letters are the multitude of created things present in the universe." (53)

Some of us, who are keen observers, may often see and hear God's creation speak to us. Or we may be able to see creatures who live alongside us trying to communicate with us about their identity and dignity as well as their love for God and His creation.

How can we listen better to the wisdom of Nature?

Many things silence their voices and harden our hearts to the extent that we lose our capacity to hear them. One thing we can do is "free ourselves from noise and busyness if we are to hear the word of love being spoken to us in the creature before us." (54)

St. Francis of Assisi with the Animals, Patron Saint of Ecology by Maria do Carmo Silva Teles; photo: Pedro do Sal

How do we connect with other creatures of God every day? I have witnessed many who love to interact with animals and even insects. When creatures of God come to graze in our backyard, I have often tried to commune with them, though it sounds a bit crazy, especially talking to them from inside my room. When I post videos of my efforts to speak with them on my Facebook page, many friends of mine would laugh and say, "They always look at you as if they can hear and understand you." I love to talk to animals while taking videos of them. It is very fascinating to see their reaction or response to my effort to reach out to them. I am convinced that animals can feel our love and care, even if we don't understand their language.

We are deeply connected with them on a very spiritual level. I have heard how St. Francis of Assisi even preached to animals when people refused to listen to his sermons. Our connectedness with other creatures of God in the Risen Christ is very evident in the teachings of Pope Francis, especially in his encyclical Laudato Si, where he says,

> "Christians, all the creatures of the material universe find their meaning in the incarnate Word, for the Son of God has incorporated in his person part of the material world, planting in it a seed of definitive transformation." (55)

God always does His part in creating and sustaining His beautiful creation. His glory is revealed in all created things, and humans' task to care for the whole universe is embedded in humans. God's creatures share the responsibility to appreciate, nurture and use His creation with love and gratitude.

Pierre Teilhard de Chardin, in his book Christianity and Evolution, argues that God:

> "makes things make themselves. That is why there is no breach or cleavage at the point at which [He] enters. The network of determinations remains intact – the harmony of organic developments continues without discord." (56)

To see and understand God's creation and His purpose and plan for us, we need to realize our own realistic, holistic spirituality, encompassing the realities of both Heaven and Earth. "Thomas Berry, one of the most prominent voices in eco-theology," warns us that to be spiritual means to be continually converted. Mainly, humans must realize how they have harmed the earth. Even Christians have set aside God's creation in their pursuit of everlasting life. They have missed that Heaven and Earth are part and parcel of Christian spirituality that can lead us into eternity. (57)

God continues to speak

Our God is the God of both Heaven and Earth. This God who created both has only one purpose, that is, to let His beloved children share His glory forever. He continues to speak to us, His beloved people, in every generation through the people He places in our lives. The letter to the Hebrews tells us:

> "In times past, God spoke in partial and various ways to our ancestors through the prophets; in these last days, he spoke to us through a son, whom he made heir of all things and through whom he created the universe." (58)

St. Paul, in his letter to the Colossians (59) tells us that Jesus indeed is "the visible image of the invisible God, the pattern of creation, the one in whom the divine came to dwell." (60) Many of us need to ask ourselves if we have forgotten or ignored God's communication with us.

Mary came to La Salette to plead for our conversion

After the French Industrial Revolution (1760-1840)— which broke down spatial barriers and accelerated economic globalization – God sent help through the Blessed Mother to Maximin and Melanie at La Salette. He sent the Blessed Mother to warn us that we need to change our way of life. All damage done to God's creation is a human fault. Human consumption and waste have impacted God's creation so much that even the Earth's produce is either poor in quality or harmful to human health.

In our day and age, we need more people to heed God's warnings. I admire many people who always try to conserve energy and consume all the food on their plates. When I eat at restaurants, I always bring the leftovers home. I am convinced that it is immoral and perhaps sinful to waste food from the hard labor of farmers and poor workers. Reckless human activities such as unequal distribution of food and waste of food have brought many people worldwide to starvation.

We must change our bad habits

There has been unequal distribution of produce and food around the world. We can challenge ourselves to consider how humans, especially Christians and the people of goodwill, can become more aware of the tragedy of irresponsible lifestyles and consumption and correct the wrong before it is too late.

Many people around the globe have become aware of drought, famine, and starvation. Notwithstanding, some would rather blame others for natural calamities. Others even dare to say that there must be no problems if God is good, loving, forgiving, and merciful. God, who is powerful and loving, must see and prevent our needless sufferings.

Sadly, many fail to hear and heed God's warnings. Jesus warned the human family against the danger of ecological problems by sending the Blessed Mother to La Salette on September 19, 1846. The Blessed Mother said, "If the harvest is spoiled, it is all your fault. I warned you last year through the famine. When people see their spoiled harvest, they cursed ..."

The Church must help in forming better protections for God's Creation

Christians and other people of goodwill are tasked with codifying better environmental systems. We must also lay out feasible plans and act on them to help us resolve our environmental ecological crisis. This has become one of the most ominous threats we all face in human history. Our awareness of it can help us look deeply into ecological issues rooted in reckless human activities.

In the history of the Catholic Church, the danger God's beautiful creation faces due to irresponsible human activities has long been warned against since the time of St. Pope Paul VI. The good saint in his apostolic letter Octogesima Adveniens (May 14, 1971), laments the sad and reckless activities of humans that destroy nature.

We are creating our own suffering and precarious future

The very thing we humans absentmindedly exploit turns us into victims of our own actions. We begin to suffer from the effects of polluted air, water, and soil. Both physical lands and the whole spectrum of natural environment become beyond human control to the extent that future generations will have irreparable ecosystem. The whole human family around the globe is affected by our ecological crisis. Christians together with the people of goodwill must take charge in repairing the damage done to God's creation. (61)

St. John Paul II also sternly warned us against the danger we will face if the ecological crisis isn't addressed by reforming our greedy human wants. In his encyclical letter Redemptoris Missio (The Mission of the Redeemer), March 4, 1979, St. John Paul II reminded us that we generally see nature as no more than its usefulness for our immediate wants and needs.

On the contrary, God's will was that we must be in touch with the Earth and everything in it by intelligently guiding and safeguarding it. We must not utilize and abuse it absentmindedly without considering its sacredness and usefulness as God had determined it. (62) In our age, many find it a need for our livelihood. Many continue to fail to see the need for communication with the whole of God's good Earth.

Why do we continue to allow the devastation of the Earth?

Without humans' realization of the damage that we have done to the environment, God's creation will continue to suffer. We always often use our God-given talents and resources to meet our seemingly insatiable needs. What is worse is our every effort to accumulate everything we can is merely to satisfy our undying desire for more material things.

Icon for Recycling to Save the Earth

The aftermath of our undisciplined, self-centered activities is often devasting for all the creatures of God and the whole of creation. The spirit of care for – and protection of –God's creation must be restored by returning to the roots of our Christian spirituality.

Ferdinand Nwaigbo, in his article "Pope Francis and the Programme of Ecology in a Time of Change," eloquently states that our greedy, materialistic, and industrialist activities cause sufferings for the natural environment. Changes in climate and natural calamities reveal how much God's creation has suffered.

God said that his creatures were to "cultivate and care for" the Earth (Genesis 2:15)

Despite God's wish that we "cultivate and care for" the Earth, one of our primary goals has become making a living at the expense of creation's health. We have used our gifts and talents to make new inventions and technological devices, even to take advantage of natural resources without conserving them for a proper, future use. Our loss of our ability to see the natural environment from a God-centered approach is the root of every crisis in the natural environment. (63) Even the moors of our Earth – those uncultivated hilly areas – have become bleak and barren from the devastating climate change and natural calamities happening regularly. Our greed is the most dangerous and urgent attitude we all need to address in our age.

During his general audience in 2001, Pope St. John Paul II called for the conversion of Christians and people of goodwill to help change the way we live to save the natural environment that provides us with basic necessities. He even reminded us of the need for conversion in conjunction with the Earth's ecosystem.

We are called to a conversion of heart, mind, and soul

Conversion as a response to the ecological crisis will help prevent future natural catastrophes. We have lost our role as the Creator's custodians who were tasked to take care of God's creation. We must stop being absentminded caretakers of creation and, instead, work to save the Earth from all forms of illness and calamities. (64)

We need to cooperate with God, who is always at work, to reconcile the whole of creation to Christ. Our collaboration with others is crucial because we have the power either to cause destruction or help save and ensure the future of God's creation. (65)

What are some ways of ensuring the future of our Earthly home?

How do we prepare ourselves for a continuing conversion experience in our lives? How helpful are our human environment and the people we live with to make conversion happen? According to St. John Paul II, family life is essential for our integral development.

First: Begin with the Human Family

He explains this well in his encyclical letter Centissimus Annus (On the Hundredth Year Anniversary of Rerum Novarum), May 1, 1991. The human family is the genesis and foundation of human ecology. It is where a person learns about "truth and goodness," what it means "to love and to be loved," and how to become a healthy human person.

The reciprocal love and care of married couples prepare and sustain a kind of environment where their children are born, raised, and molded to meet their potentials. Every child's human dignity is nurtured, and each person is prepared to reach his distinct destiny. (66) Right relationships matter in the spiritual life. They can have a life-changing experience and grow spiritually.

The family can teach every family member to participate in the protection of God's creation by changing their outlook on consumption. They can conserve energy, reduce unnecessary accumulation of material things, and become more resourceful. A change of heart in everyone is indispensable, although it is always challenging. Every family, in this way, helps address the challenges of both "the institutional and personal dimensions of ecological spirituality." (67)

Pope Francis warns us how dangerous it is to buy and spend unnecessarily. We must be aware of the hidden dangers of the ongoing and excessive accumulation and use of commodities. The fact is that "since the market tends to promote extreme consumerism in efforts to sell its products, people can easily get caught up in a whirlwind of needless buying and spending." (68)

Second: Speak and act to protect the Earth

We sometimes forget the power of our voices and actions in the fight against the onslaught of consumerism. If we are not careful, we all can be trapped in the web of endless and unsatisfying gratification with material things. The impact everyone and every family can make in the journey towards a healthy ecology is power against socio-political and economic powers.

Pope Francis encourages us to change our way of living because our healthy lifestyle can influence those who have socioeconomic and political power. Consumers can enormously influence on economics and politicians by expressing their disapproval and dislike of certain products.

Consequently, business owners and makers will consider the environment's needs, and those of the consumers. (69) We have contributed to the change in human society thus far and have tried to become catalysts for change in human behaviors matters in the spiritual life, yet much more change is needed.

Authentic and lasting change must include attitudinal, behavioral, and value changes

To impact change both in our lives and in the lives of others, we must take care of many essential components. Ferdinand Nwaigbo argues that only attitudinal, behavioral, and value changes constitute authentic change. Christians' identification and relationship with creation make change happen on a personal and communal level. A person's ability and capacity to change requires tenacious faith and resolute practice of faith. (70)

We all have different spiritualities due to our upbringings, cultural mandates, and relational necessities. Even then, we still say that it is easier said than done. Nevertheless, we who believe in the Risen Christ who saved the whole of the universe must understand and practice an ecological spirituality that emphasizes "the human initiative in caring for God's creation." (71) Who can learn about and teach the importance of caring for God's creation?

Our family can create and instill basic gospel values for life

The family has the capability and responsibility to create an environment where humans and nature are protected and nurtured. The truth is that everything begins and grows in the family. This truth is always explicitly taught by the Church in her official documents. For example, the Compendium on the New Evangelization, no. 777, reiterates the Church's teaching on the crucial role of the family in teaching children human and gospel values. It reminds us that in the family, we learn every basic and essential thing in life – praying, working, saying nice/good things, loving, forgiving, reconciling, and living daily life to the best of our abilities. Our parents teach us almost everything we need to know. (72) What values and how much have we learned from our families concerning environmental protection and human activities that impact God's creation? I am sure that we all can relearn the pressing issues of our time, particularly about human activities and ecological crises that arise from them.

With the understanding of our own identity and destiny given to us by our respective families, we can participate in the care for God's creation. The late Pope Emeritus Benedict XVI, in his speech to the clergy of the Diocese of Bolzano-Bressanone, revitalized the essential connection between the natural environment and human activities. He explained how human activities have impacted God's creation and how Christians must bear witness to the truth about their faith in the God who provided creation and redemption for the earthly life and the eternal life, respectively. He pointed out the connection between faith and actions as a sign of earthly and eternal life.

God's creation and redemption are to be put side by side because we cannot live our lives here and hereafter without one or the other. Our Christian responsibility is to involve ourselves with conviction and resoluteness in the views prevalent among the general public. We must set examples by the way we live our lives.

We must spread a message that is based on our own beliefs and that is viable for others to practice. We must rely on the Lord's help in living out our faith and becoming credible witnesses to our faith in Jesus in this age (73). It is a great challenge for most of us to become faithful and credible witnesses in this age!

Competition between commercial and family-scale goods

Commercial products are done exclusively for the massive production of goods from the plantations and affect the family plantations. Poor farmers have no choice but to compete with the demands of multinational corporations. I recall the response of my relatives and friends to my question about the disappearance of edible beans.

These edible beans, which were very popular in Burmese homes when I was growing up, can be boiled and eaten with any meal – breakfast, lunch, and supper. They are so rich in protein that vegetarians rely on these beans. Even my father, a vegetarian, loves to have them for his meals once a week. I was very sad to hear them say that they no longer plant those beans because there are very few or no more demands in the market. It is worse to learn that they have started using pesticides, often excessively, to meet the needs of the trade market. The long-term damage caused by this reality is explained well by the Church.

We take care of ourselves when we take care of our world

Pope Francis, in his encyclical Fratelli Tutti, warns us how much God's creation has been suffering and is in dire need of restoration. We take care of ourselves when we take care of the world we live in. We need to think of the one family of God; many families inhabit a home common to all.

Profit-oriented economic platforms and systems may never care about creation. Those who have socio-economic powers constantly ridicule and silence the responsible voices of those who care for creation. The cultures of our times – of little depth – challenge us to stand up against systems and practices that deplete specific resources. If not preserved and taken care of, the depletion and loss of natural resources will lead to more conflicts, violence, and wars, even in the concealment of any honorable end. (74)

Polluted air, water, and earth could lead people to wage war against others to secure basic needs. Therefore, very conscientious person's response to the needs of creation is of paramount importance.

Seeking a balanced spirituality of involvement

Believers have a responsibility to respond to the needs of others, creation, and themselves. This responsibility reflects their spirituality, be it one-sided or balanced. A balanced spirituality always responds to the needs of every part of God's creation. Bernard Lonergan argues that "the language of spirituality" is "a way of answering needs" of every created reality. As a concrete sign of "religious conversion," one would engage in "four distinct levels of human activity: experiencing, understanding, judging, and deciding." (75)

The members of the Adult Class on Aquatic Ecology go a nearby river to practice what they preach

What must be our answers to creation's needs? What are our conscious experiences? How do we understand and judge them and make decisions about them? We need a daily conversion to see the needs of creation and respond to them accordingly.

Christopher Vogt speaks of solidarity as a solution to the crisis in the ecosystem. The good of both the individual and the community that is interwoven must be balanced. When humans show solidarity with the rest of creation, they live a morally responsible life that respects every creature of God. They then learn to experience nature and cultivate a spirituality that connects them with the rest of creation. They are also able to do socio-economic and political actions that "build up and protect the common good." (76)

We live in a society where individualistic lifestyles are on the rise and the common good is being set aside. Christians and people of goodwill have both a responsibility and duty to work for the common good and the good of creation. Both must be safeguarded at all costs for the benefit of all.

St. Paul, in his letter to the Romans, speaks of total redemption for the whole of God's creation. Every creature and the whole of creation await the redemption Jesus Christ has wrought. He highlights the comparison between the two realities – the trials and tribulations in this life and the triumphs and victories in the next life – because of the Risen Christ.

Even the whole of creation, he believes and construes, will be saved together with God's children. God's creation and people share the same fate and long for total redemption. The Holy Spirit groans with us and for us, giving us undying hope for the total freedom Christ promised. As we all have hope because of Christ, so does creation because of us. (77) We must care for God's creation the way we see and care for us.

Questions for Reflection:

1. How do we see God's creation?
2. How do we participate in caring for God's creation and share its longing for redemption?
3. What measures have we taken to care for God's creation individually and collectively?

Chapter Four

The Necessity of Conversion

"Repent, therefore, and be converted,
that your sins may be wiped away,
and that the Lord may grant you
times of refreshment and send you
the Messiah already appointed for you,
Jesus, whom heaven must receive
until the times of universal restoration
of which God spoke through the mouth
of his holy prophets from of old.
(Acts 3:19-21)

Lord, guide me
along the path of conversion.
May I always realize
how much I need Your grace in redirecting
my life along the path that
You have traced out for me.
Renew the true dignity and identity
which you gave me.
I always trust in Your promise and guidance
every day of my life.

La Salette ministry in Myanmar can be symbolized by the dome of a Bagan Buddhist Temple together with the La Salette Cross

Human Frailty and the Need for Redirection

We humans are limited beings with flaws and weaknesses. We often make mistakes even when we try our best to be our best version. St. Paul spoke about the nature of sin and grace and the battle of the two within him. His reasoned and holy desires were constantly confronted and overcome by his inclinations to sin and his contradictory actions. Only the power of Jesus' saving grace could help him overcome everything contrary to his identity and dignity as a faithful follower of Jesus (Romans 7:13-25).

There is a belief that even the holiest saints make mistakes and commit seven venial sins daily. So, the need for everyone to convert is not overstated. The questions we may have are what can involve conversion and how hard it might be to undergo the conversion process. On the one hand, conversion involves risks, detachments, and sacrifices. On the other hand, it benefits both the converted and their relationships, surroundings, and environments they are in. For these reasons, our understanding of conversion will encourage us to take concrete steps and actions to strive for the common good. Each of us will differ in our approach to it because we all do things following the level of our understanding.

Of all definitions of conversion, Patrick Madrid's understanding of conversion is the most convincing when it comes to the need for sacrifices every human person with a good moral conscience must make daily to protect God's creation.

He concurs that it is the truth that enables us to make sacrifices in daily life. When we decide to convert into something better, we become martyrs in many ways. We must let go of something or someone we love or are accustomed to. It takes our whole being and all five senses for Jesus Christ. We become docile and willing to let the truth lead us. For most of us, it may mean doing things we would instead not do or going to places we prefer not to go. It is just like Jesus reminding Peter to prepare for the mission he might be sent to (John 21:18-19). (78)

Our conversion can affect many others

Our conversion can affect the lives of others. Those living with us, especially our loved ones, experience the immediate effects of our conversion because they can see and witness changes in our lives. One of the things that we all need to undergo for conversion is the ability to look at other people with an open or welcoming attitude.

Joseph Ratzinger, in his book Christianity and the Crisis of Cultures, argues that even our decision to look at someone face-to-face is a decision to undergo conversion. In such a decision, we go beyond ourselves and make room for that person, so much so that we will have a mutual encounter with each other. It is a testimonial nature of a moral value that largely depends on our free and covert resolution. Our determination to see someone evoke a change in my life. (79) Hopefully, how we look at creation, others, and ourselves changes for the better every day so that we become beacons of hope for creation.

To convince ourselves that we are already forgiven and living in God's grace, we need to have a conversion experience, just like St. Paul, St. Ignatius of Loyola, and even John Wesley. They all experienced God's grace, which made their conversions possible.

Brendan Byrne, in his journal "Ignatius Loyola and John Wesley: Experience and Strategies of Conversion," explains that the conversion of John Wesley first and foremost depended upon God's enabling grace as a priori. His human response corresponds to God's initial and sustaining grace that helped him go through the conversion process. His conversion enabled him to repent and reform his life. (80)

Conversion is often an option

The experiences of saints and sages of every age teach us the possibility and reality of conversion in response to God's call and the needs of others. We need to be converted because the urgency of creation's needs can no longer be ignored, let alone sidelined in our daily activities.

We may want to look for our need to be converted, especially when we feel we have been faithful to our religious, moral, and relational duties. St. Pope John Paul II, in his encyclical *Solicitudo Rei Socilais (On Social Concerns)*, teaches us that the importance and reason for our conversion necessitates behavioral and mental change because we can change either or both of them when we undergo conversion.

Conversion presupposes our relationship with God, and with all sins we have committed and their consequential effects on our neighbors and communities. (81) It is related to the common good and includes all of God's creation. To be converted is imperative for all Christians. I am convinced that how often we talk about conversion in our ordinary conversations matters the most. Furthermore, our continuing efforts to teach others about it, particularly to younger generations, is also of great importance. These activities constitute our own conversion experiences because we only share willingly and faithfully what we ourselves have experienced.

Conversion and younger generations

Cinderella Sequeira, in her article "Laudato Si': Caring Beyond Limits With A Cosmocentric World Vision," reminds us that younger generations are born into the digital age with technological devices in their hands. Our limitations and failures to bring them up in an environment where they interact more with others have forced them to grow in the virtual world where only the mind controls everything. Sadly, the creation of human hands, especially the virtual world, morally blinds us "to the destruction of our wildlife, flora, and fauna." (82)

I am here reminded of a La Salette Missionary, Fr. John Gabriel, M.S., who loved to say, "Technology is good when it works," and I would add to his remark, "for the wellbeing of every human person." Professionals use materials from creation to make electronic devices. The amount of benefit returned to the natural environment determines the morally usefulness of technology. Otherwise, we must change the way we utilize technological and electronic devices. It means that we must change our way of living.

Creation provides for us

I remember my grandfather teaching me to change my way of life as a teenager by changing how I looked at others, especially those who could not defend themselves. He would specifically talk about God's creation, which needs our protection.

One time, I tried to capture a reptile. My grandfather told me not to do it unless I was willing to eat it. He taught me the same thing in using other parts of creation. For example, he would keep all pieces of teak wood in the attic of his kitchen. I remember how lovingly and gratefully my uncle used those teak wood pieces for a few years, even after my grandfather's passing. My grandparents always reminded us, their grandchildren, that it is from creation, or nature in their expression, that we have all the basic needs – food, clean water, fresh air, and decent shelter.

They also reminded me to be courageous in making wise choices and tough decisions in daily life. Though Buddhist, my grandparents seem to have a similar mindset and attitude toward God's creation as Pope Francis. During his General Audience in September of 2021, Pope Francis said, "Let us pray that we all will make courageous choices, the choices necessary for a simple and environmentally sustainable lifestyle, taking inspiration from our young people who are resolutely committed to this." (83)

The place of the scriptures in our conversion process

There are many ways God speaks to us every day. God also uses the events of life to help return to Him. Family fortunes or tragedies can move people to convert to God. But what affects our conversion the most? Who motivates us to change our ways of thinking and living? How does one undergo the process of conversion?

To answer these questions, we need to look into the role of the Sacred Scriptures, for they can communicate to us that Jesus Christ was revealed as the Incarnate Son of God. Sacred Scriptures are the norms for our Christian way of life. To grow, nourish, and sustain itself, Christians' spiritual lives need the Sacred Scriptures as their foundation. (84)

Reading "the Book" can change our lives

We need to read the Scriptures with love and devotion to find answers to life's questions, some of which are stated above. The Dogmatic Constitution, Dei Vervum (On Divine Revelation) from Vatican II reminds us of the importance of reading the Bible with love and devotion. Careful preparation and prayerful reading of the Scriptures are paramount for every devout Christian. The more we read it, the more we can know and love the person of Jesus Christ about whom the Sacred Scriptures teach.

The celebration of sacramentals, sacraments, and other spiritual practices can help us to understand and love the Sacred Scriptures and receive the grace of God in Jesus Christ by the power of the Holy Spirit. Our prayerful reading of the scriptures engages us in the conversation with God and helps us to encounter Him. God speaks to and hears us in our prayers (85). I believe that reading the Scriptures is having a very important conversation with God face-to- face. Reading the Scriptures can help us learn to love God and live in His grace daily.

The Word of God, the Sacred Scriptures, is the best guide to teach us everything, especially our faith and spirituality. (86) God communicates with us through His Word, the Scriptures. Human words are used to express, although inadequately, God's Word. We see how human words help us understand and express God's Word. The more we read the Word of God, the more we learn to express ourselves better as God's children. If we notice it, we use many different words daily to be able to effectively communicate our feelings, thoughts, dreams, and plans.

We are called to be Christ's witnesses by our lives of faith

Frank P. DeSiano, in his book The Seven Commandments of Discipleship, explains how God molds us into His people by shaping our minds and the words we use with His Word. God forms human words and weaves them into a river of human language through His actions and relationship with humans, as evidenced by the Sacred Scriptures. God also fashions the minds of His believers following His design. (87) So we must allow the Word of God to transform us according to God's design.

Since God's Word shapes us in many ways, we are changed in accordance with the level of exposure and our response to God's Word. DeSiano claims that "because God alone is God, God's Word alone has a place of absolute importance in our lives." (88) It may not sound real for some, but for believers, God's Word is an absolute necessity for spiritual growth and maturity. It can change our lives around because God's Word has the power to transform us so that we change the direction of our life in everything we feel, think, say, and do. The Word of God can mold us until the day we become His credible witnesses. (89) We who believe in God and read His Word (Sacred

Scriptures) know that Catholic spirituality is rooted in the Word of God. Catholics practice Catholic spirituality to grow in the way God calls them to.

We exercise spiritual practices rooted in our Catholic spirituality to grow in God's image and likeness (90) or reshape it if we have disfigured it. Our practice of Catholic spirituality with great faith helps us in our efforts to transform ourselves every day, that is, to undergo a continuing conversion of our hearts until we become who we are created to be. Conversion helps us regain our original self, which is our true self.

We are all aware of how things distract us all the time. Consequently, we are disoriented in many ways every day. To navigate through the sea of distractions or free us from the tricks and traps of distractions, we need to ask these questions:

- Which part of spirituality do we need to practice now when we face an environmental crisis that endangers all species on this planet?
- What kind of change do we need?
- And where do we begin?

We must live a new way of life to live a healthy and holy life with the rest of creation.

Are we participating in a new way of living on Mother Earth?

Konrad Raiser, in his book Challenges and Hopes for a New Millenium to be the Church, Risk Book Series, contends that we must see "conversion as a Way of life in Cultures of Violence" and that our lives must also change so as to contribute to the systemic change in "the cultural consciousness." The essence of conversion is geared towards the change in structures or systems in human society. A genuine conversion is "not a momentary act of moral decision, but a process of learning and a new way of living." (91) o help us realize the way we live our lives, we need to listen to competent authorities. Pope Francis often speaks of a "throw-away culture" to highlight the importance of the need for our conversion that helps change the lives of many that greatly impact ecological issues.

Arthur Darby Nock, a professor of the history of religion at Harvard University, says that converting means directing a human soul from lukewarmness to conscientiousness to the extent that a fundamental change occurs in a person's life. It is making a choice that corrects an old life pattern and guides a new one as guided by true religious teachings. (92) Religions teach us about life and our choices. Their rituals and rules, rooted in the teachings of their founders, point to some things beyond us. These regulations and rituals instill in our minds and hearts that we have many options in life, but we must choose the good or even the best ones.

We have a plethora of choices to make

Frank DeSiano, in his book Why Not Consider Becoming a Catholic, argues that our world today offers many options for us to select. We can choose freely any produce, go to any store, consume any food, and have any career we may like. Once humans lose their rights and choices in life, they start to revolt against those who take away their rights to choose whatever they think is the best for them. (93) There are external and internal rebellions in our lives when our fundamental and inalienable human rights are taken away. We can speak for the groaning environment whose voice has been ignored, denied, and rejected. We must make the best choices for our ailing environment.

In our times, the most needed choice is the conversion of our hearts! What is significant is our willingness and actions to change our attitude and lifestyle because things constantly change, requiring us to change alongside them, preferably for the better. We can think of religious practices and political and socio-economic strategies that continually beg for changes in their approaches to sustain the old members and to welcome the new ones to the family, the church, and society at large. Seeing the need for change in our spiritual life, we are challenged to ask: Is conversion applied only to non-believers, or does it apply to all?

For Marcel Schlewer, a La Salette Missionary, reminds us that "conversion applies not only to pagans who became Christians; it is the work of a lifetime." The heart must be converted for actions to be radical and sincere without possibly falling back to the same pattern of sins. The change of heart means relying on God's grace with a total trust and working hard to allow the desire of God to influence our daily circumstances. (94)

Do our choices help to change our lives and that of others?

Only with a change of heart comes a change of behavior in the person because our heart determines the choices we make every day. The heart is "the profound source of one's free choices, the principle shaping one's attitude towards God and other people." (95) We are called and challenged to make choices that benefit all, including God's creation. Our choices must come from the conversion of our hearts shaped by the Word of God.

How involved is our personal call to Christian discipleship?

How often do we have to be converted? St. John Paul II, in his Post-Synodal Apostolic Exhortation Familiaris Consortio, reminds us that our conversion is true and becomes permanent if we embrace two realities. First, we must detach ourselves from the evil and attach ourselves to the good to progress spiritually. The dynamism of change in our lives enables us to integrate God's gifts and graces into our personal and communal lives. Our growth process leads us toward an integral transformation and understanding of Christ's mystery that we have received since the day of Baptism (96). Christ can empower us in our battle against the Evil One and any accomplices as we strive to grow as God's children and responsible and credible followers of Jesus.

The essential, not the peripheral, change must occur in the person (97) to become a credible disciple of Jesus. A disciple is credible in witnessing the teachings of Jesus in everyday living if he/she is a person of action. A twenty-first-century mystic, Thomas Merton, enlightens us and says: "Christian discipleship entails action – loving action which manifests Christ in the world, allowing the Spirit to dwell in and work in us." (98) Our cooperation with God's grace and collaboration with others indeed help to change the lives of others and be changed by others.

How personally involved are we in collaborating with others to benefit God's creation?

In the Acts of the Apostles (2:1-41; 17:1-34), we see how some were converted to Christianity through personal conversion experiences through the example, preaching and miracles done by the Apostles or, in the case of St. Paul, by Christ himself.

The signs of our times provide us with different experiences of God speaking to us so that we may be converted. One of those signs is the miseries people undergo and the damage done to God's creation. So how does God's creation speak to us in these times in human history?

The Gospel of Luke's three parables: the Lost Sheep, the Lost Coin, and the Prodigal Son (15:1-32), illustrate the particular concern of Jesus for the lost and the love of God for sinners who truly repent. The lost sheep can only cry for help, just like many lost souls in our world today. There are many who are like the lost coin with limited or no capacity to know where they are or cry for help. Many are out there whose return to the Father is patiently awaited.

Questions for reflection:

1. Are we the lost sheep who need to be rescued from our hopeless, desperate, or dangerous situations?
2. Are we the lost coin that needs to be retrieved because we can neither try to get out of the situation by ourselves nor cry for help?
3. Are we like the prodigal son, who can only be waited for because we need time to realize our own sins and make a decision to return to the Father?

Interior of the Basilica on the Holy Mountain of La Salette during a Eucharistic Celebration

Chapter Five

The Common Good

"Love one another, as I have loved you,
so you also should love one another.
This is how all will know that
you are my disciples,
if you have love for one another"
(John 13:34-35)

Lord, help me to see You in others
and the whole of creation.
May I love and serve You
in others and your beautiful creation
so that I may be saved
together with everyone in the world
and all you have created.

We must protect our fragile coral reefs like those in the Gulf of Eilat on the northern tip of the Red Sea and elsewhere

Our responsibility to protect our ailing planet

The most concrete way to show our love for one another is by taking care of one another. So, it is logical to say that to love God's creation concretely means to take care of it as humanly as possible. Concepts become a reality when they are internalized, and the person who accepts them fosters their spirit.

A bishop from Myanmar once shared his experience of doing a mission in the United States before becoming a bishop. He said how he was impressed by the Americans' concern for the common good, which is very much evident in the way the country is built and run. He and I agreed that the Americans have a Christian spirit.

When I shared the remarks of the bishop with my American friends, one of them said that the British nationals even called Americans a religious people. My own experience of Americans' sense of the common good and their concern for others is remarkable. When I first learned to drive in Atlanta, Georgia, I always witnessed drivers giving way to me as opposed to some reckless drivers who would not care about others. One more thing about their concern for others is how they let other shoppers go first by saying "after you." Most Myanmar nationals are impressed by such a gesture and expression.

My experiences throughout my ten-year stay in the United States always make me think of Myanmar and other poor countries that are left far behind developed nations. The developing nations are left behind in everything, let alone in the sense of the common good and care for God's creation.

Acquiring a sense of common good in developing nations

One of the main reasons for abject poverty and scarcity of resources in developing nations is the lack of a sense of the common good in the general populace, especially in the leaders of underdeveloped countries. If citizens, especially their leaders, are indifferent to the common good, let us remember that nationwide progress will probably be very slow. Sadly, there are no signs of development in people's livelihood and the nation. Incompetent, corrupt, and unjust leaders always keep their fellow citizens in poor living conditions. The people they are supposed to serve are constantly deprived of basic human needs and rights, so much so that they do not have time, energy, and resources to see and resist injustice and fight for their people's inalienable human rights. To survive in the poor living conditions, they also have a limited capacity to work for the common good. Their attitude and behavior can also cause them to be indifferent toward the common good of their people.

Supporting human rights and the protection of God's creation

Our human indifference to the common good has not been properly and fully addressed despite the Church's consistent teachings on the common good. As members of God's family called the Church, we are tasked with things to do for others and for God. For this very reason, it is good to remember one of our human and Christian responsibilities — to take care of God's creation.

If we recall, stewardship is included in our baptism in the Lord for we promise to follow and imitate Jesus Christ, who loves and saves all of His creation. We must help the Lord and His Church to address life's issues, one of which is that God's creation is facing in our days: air, water, and land pollution.

The important place of Catholic Social Teaching in daily life

To address environmental issues, we do not have to look anywhere else. Holy Mother the Church has profound, relevant, and effective teachings on our shared responsibility to find solutions for any problems or issues the whole human family worldwide faces. Many, even Catholics, often fail to see the meaning and implications of the Catholic Church's Social Teaching.

Caitlin Morneau, in her book Harm, Healing, and Human Dignity, jogs our memories concerning the Church's teaching on social concerns. God bestows upon every human person inviolable dignity that humans must exercise in their relationship with others, beginning in the family, the local and international communities where they are members. That is the only way to manifest the love of Jesus that is alive in every person, relationship, and community. (99)

When we think and speak of the family or community, we realize that we live with others. The reality of family/community life revolves around rights and responsibilities. These inherent rights and duties can be ignored, neglected, or sidelined. For instance, in some countries, human rights focus more on rights than responsibilities, while others do not pay attention to them.

As equally created beings, we all have both inalienable rights and fundamental responsibilities. Based on these facts, it is always timely to be awakened by the need to reevaluation of our views, perspectives, ideas, and lifestyles that we have been comfortable with. That is the only first and surest step to our active participation in the work for the common good.

What does it mean to be created "in the image of God"?

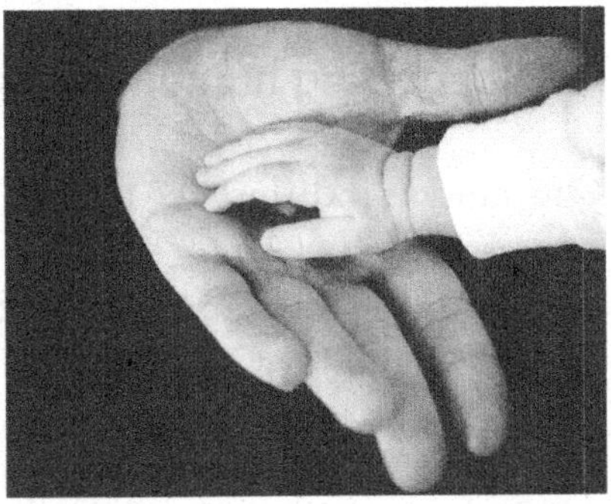

Caitlin Morneau expresses her understanding of the rights and responsibilities of humans, especially of Christians/Catholics and other people of goodwill. The interrelatedness between humans and the rest of creation is due to the fact that humans are created in God's image and likeness.

For this very reason, humans have both inalienable human rights as well as duties and responsibilities towards God, themselves, and the rest of creation. Relying on and benefiting from everything God has created, humans must preserve, protect, and support the lives of other created things. Having their basic human rights, they must also see that humans have their personal God-given rights and inherent and corresponding responsibilities towards others, communities, and the good of all. Humans must enjoy their rights and duties that are morally imperative and spiritually inherent. (100)

A poem by Mary Oliver to reflect on . . .

Knowing and understanding God's creation is the initial and essential step for our genuine conversion that alone will convince us to care for God's creation. Before going into a serious exploration of the need for our responsibility to care for God's creation, let us contemplate the beauty of God's creation and its place in our lives and in the world through the lenses of Mary Oliver's poem Sleeping in the Forest. I include the exact words of her poem to help us focus, contemplate, and understand better the reality and necessity of God's creation in our lives. Every word she uses gives us a vivid picture of how nature operates its ecosystem, with or without our knowledge and awareness.

> I thought the earth remembered me, she
> took me back so tenderly, arranging
> her dark skirts, her pockets
> full of lichens and seeds. I slept
> as never before, a stone
> on the riverbed, nothing
> between me and the white fire of the stars
> but my thoughts, and they floated
> light as moths among the branches
> of the perfect trees. All night

I heard the small kingdoms breathing
around me, the insects, and the birds
who do their work in the darkness. All night
I rose and fell, as if in water, grappling
with a luminous doom. By morning
I had vanished at least a dozen times
into something better.

Her experience of nature doing its job well not only inspires us but fires us up to love and nurture God's creations as humanly as possible. Her expression of her love for nature based on her own experience may touch our hearts so much that we will savor every moment of our lives in union with the whole of creation every day.

She tells the story of her experience of nature reconnecting with her with tenderness. Nature is described as a woman putting her black dresses in order. She (nature) has her pockets with flaxen fabrics and nuts. Insects fly around among "branches of the perfect trees." All insects and animals build and stroll around their little domains. Her experience of being in nature's welcoming environment made her disappear into a better world more than twelve times. (101)

Anything poetic may inspire and change everyone, including some people who are not fond of poetry. Poetry speaks to a person's heart and can touch the core of everything one believes and loves. The questions we may ask here may be: How far are forests still teemed with creatures of God? How peaceful and tender are they in our times? Or we may still hear the cries of creatures and the whole of creation if we listen to them on dark silent nights.

The Church reminds us that the whole of creation has suffered due to unproportionate human activities. Every human being is challenged to change their outlook on life and strive to change their lifestyles to be able to participate in the protection of God's creation actively.

Our call to be stewards of God's handiwork

Being aware of the needs of creation and having heard its cries through professionals and experts in ecological crises, Pope Francis teaches us that our "ecological crisis is also a summons to profound interior conversion." Therefore, "living our vocation to be protectors of God's handiwork is essential to a life of virtue." (102) We must learn to live with gratitude towards our Creator and thank God for the gift of creation and become aware of our connection with the whole of creation, and take responsibility for it seriously. (103)

We all are called to become God's credible and faithful stewards of creation. God has given us enough resources and graces to become faithful caretakers of His creation. We can claim that even people of faith and of goodwill, no matter how few they may be, can set the world on fire to protect God's creation. Many can become missionaries of God's creation.

There are many reasons for such a missionary endeavor. Many of the reasons can be found in the writings of professionals. For instance, Xue Jiao Zhang, in his article "How St. Francis Influenced Pope Francis' Laudato Si," reminds us that "Mother Earth not only accommodates us but also provides food and resources that we need in order to make a living." (104)

Let us understand, first and foremost, what creation really is and who we can connect with through it. Matthew Fox, having understood creation as one of God's Original Blessings, tells us that creation is "a trace, a footprint, and offspring of the Godhead. Creation is the passing by of divinity in the form of isness." (105)

What happens when humans fail to protect but damage creation? We see the consequences of human behaviors right before our eyes, whether we realize it or not. It is not only morally wrong but also spiritually ill. Nelson Kloosterman warns us that humans abominate Mother Nature by committing sins against her. The rise of sins against ecology is due to the duplexity mentality that exploits the misuse of rich natural resources and living things as disposable items at human whims. (106) So, we see more reasons to understand creation and eventually be converted to care for it, for it is God's original gift to us daily.

What are the "twin laws" which guide us?

People who are truly converted feel, think, speak, and act in accordance with twin laws that inherently bind us. These twin laws are natural law, which "deals with ethical dimensions of responding to those in need, and divine law, that expresses the Creator's will for all people." (107)

Both divine and natural laws tell us that we coexist with the whole created world of God. This necessitates our every effort to care for it because every part of creation affects every other part of it, which includes all of us, the human species.

There is a human tendency to disconnect ourselves from the rest of creation unless and until we find its usefulness for our needs or greed. Living in the age of constant connection with the whole of humanity through the use of information technology, we are challenged to explore and understand the connectedness of all created realities. What might happen to human connections and the connection between the Creator and His creation is both exhilarating and fearful. We must never forget our identity as creatures of God endowed with the ability to connect with others and the rest of creation. When a part is separated from the whole, the separated part may suffer, wither, and die. Therefore, to live, especially to live our potential to the fullest, we need to stay connected to the whole of creation and our Creator.

We are all part of the whole of creation, which yearns for our care

Ronald Rolheiser, in his book The Holy Longing: The Search for a Christian Spirituality, explains that everyone is a part of the whole of creation. Only if we see ourselves in the milieu of the whole universe, can we attain the proper comprehension of ourselves and the essence of spirituality. (108) We must be able to see the longings of others and the whole of creation, just as we ourselves have our own. Rolheiser remarks that "we are part of a universe, that part that has become conscious of itself, wherein everything yearns for something beyond just itself." (109) Just as we all have insatiable yearning, the universe has its own. The creation of God is yearning for our care and protection.

The understanding of humans' efforts to understand themselves may differ from one school of thought to another or from one religious tradition to another. According to Daniel Minch, Christians' understanding of themselves is mainly reflected by how they see, understand, and experience God and His creation. Their view of their Creator and the creation also impacts how they live their lives and relationships with others in society. (110)

We can become closer to God through our ordinary daily living

Concerning the encounter with God in concrete human affairs, Edward Schillebeeckx highlights the importance of ordinary and mundane things, such as Jesus' interactions with ordinary folks, to let them experience freedom and redemption. In and through such ordinary things Jesus shared with the people, He brought about the closeness of God to His people in concrete circumstances. Jesus revealed and made the encounter between God and His people possible in space and time. Jesus enabled people to understand and experience what used to be obscure and incomprehensible. These acts of Jesus brought liberation to the lives of people. (111)

Since creation is originally good and humans benefit from its goodness, humans are free to use it for their good and are responsible for the sustenance of creation's original goodness. Our care for creation manifests our love for God because He is present in creation. Humans can know God through creation, which always manifests God's active works. (112)

What we know influences how proactive we are regarding the care for creation. For this very reason, we must first understand the original meaning of technical terms in the context of their respective fields of studies. For example, Schillebeeckx defines:

Louisa May Alcott, the American novelist, once wrote: "Every house needs a grandmother in it."

> "secularism as an ideological, anti-religious program of rational control of the world through technological and political means, while secular or secularity simply means finitude, that which is not godly." (113)

Living in a world that is religiously, economically, politically pluralistic, and polarized, it is not easy for many to be aware of the damage humans have done to God's creation. Human's too much focus on mass production for prosperity, even at the expense of the degradation of creation, has done a lot of damage to every created reality and being.

Richard Leonard, in his book Where the Hell is God, reminds us that "the earth is much more fragile, and more balanced and finely tuned than we thought. It is evolving all the time, now, groaning in [the] act of giving birth." (114)

Our reliance on God's guidance

It is indispensable for us to know, Richard Leonard construes, that the way we acquaint with our finiteness constitutes our comprehension of the whole of the universe. Physical realities transcend themselves and point to their genesis, God Himself. (115)

In and out of season, God keeps reminding us that humans' graspingness and acquisitiveness force people to make poor decisions regarding atmospheric conditions and climate-related issues that constantly confront us. (116) If only we all realize how powerful human actions can be, both positively and negatively, we all will surely make better, even the best decisions in life, especially when taking care of God's creation. It is noteworthy to remind ourselves often that "every human act is also an experience of the absolute limit as well as an act of transcending relative limits – becoming more than we were previously." (117)

It is mistaken to believe that we can exist independently

The faulty mindset of our time is that each human person or species can exist independently. The exclusivist attitude contributes to the ongoing erosion of community spirit or common good. Either to grow or regress, all species benefit or are victimized. Pope Francis profoundly said that neither the natural environment nor the human environment degenerates alone. They both decay together. It is necessary to look into the reasons why human and social environments decay and heal the illnesses they suffer from. The poorest and the powerless suffer the consequences of destructive human decisions more than the rest of society. (118)

Without proper and timely realization of our interconnectedness with all other creatures and the rest of God's creation, we fall prey to the devasting dangers of environmental degradation. Even Pope St. Paul VI forewarned us about this danger in the nineteen seventies, "Due to an ill-considered exploitation of nature, humanity runs the risk of destroying it and becoming, in turn, a victim of this degradation." (119)

Like his predecessors, Pope St. John Paul II also foresaw the dangers human activities pose by living reckless lifestyles. Humans are accustomed only to instantly utilizing and consuming everything nature provides. They see nothing in creation beyond its usefulness. (120) Human, economic, spiritual, and environmental growth are all interwoven.

Pope Emeritus Benedict XVI's remark sounds loud and clear. "(We are called to) eliminate the structural causes of the dysfunctions of the world economy and correct models of growth which have proved incapable of ensuring respect for the environment." (121) As every generation adopts its own culture, every life on earth is affected. With changes in cultures come changes, even devastating changes, in the rest of God's creation. Yes, it is true that nature deteriorates in connection with the cultures that fashion the synchronized human existence. (122)

We should realize that Earth is our common home

The leaders of several faith traditions met for the Interfaith Climate and Energy Conference in Israel to discuss our owning of and response to the ecology crisis

With the rise of Pope Francis came the response to the call of God to protect His creation. He fondly calls Earth 'our common home.' He calls on every believer and the people of goodwill to join forces to protect one of God's best gifts to all creatures of God. According to him, we all must understand that the most urgent and best way to protect our 'common home' is to work together in finding viable and holistic development. Only then shall we be abreast of the positive sides of the reality of change. (123)

Our rapidly changing world presents both challenges and opportunities. We must come together as one human family to care for our ailing home. We are encouraged by the fact that no nation on earth can solve the problems alone. Every human community must combine to have a maintainable and varied agriculture. This method will help us to create recyclable and less contaminating but potent energy that helps to control and make the best use of aquatic and woods resources. It will then guarantee common accessibility to safe drinking water. (124)

Every aspect of life is crucial in building a better world where every one of God's creations is loved, cherished, and taken care of. For these reasons, Pope Francis teaches that economic principles must not control political systems. Similarly, scientific and technical knowledge experts must not be the only people making decisions on economic theories and practices. The common good in our days demands that those who work in political and economic fields have open and honest conversations. Only then will policies, programs, and developments not be detrimental to human and environmental life but at the service of the precious life God has given to all created beings and their environment. (125)

Our call to be agents of change

Whether we like it or not, human life revolves around many undeniable factors that build the lives of very few but destroy more lives. The sad reality is that many compete for domination; and consequently, their failure to collaborate destroy the lives and livelihoods of many of God's creations.

Despite the damage done to creation, it is not too late to reflect and act on the realities of economic systems and become agents of change. We must remember to confront real economic issues with robustness. Our participation in economic policymaking will help all companies and businesses run well and provide good opportunities for the best income-generating jobs. (126)

What is the best we can do to transform the systems that do not benefit everyone but a few only? We must admit that it is none other than a certain change for everyone. Pope Benedict XVI taught that "a change in lifestyle could bring healthy pressure to bear on those who wield political, economic and social power." (127) An excellent concern for others on the part of those with socio-economic and political power will create a better collaboration among the powerful and the powerless as well as between the employer and the employee. The common good is attainable if everyone in society comes together as one family. The lives of many will change consequently and eventually.

Perhaps a radical lifestyle check is necessary

How many of us contemplate the change in our lifestyles? I would bet many are unaware of the need for change in their lives. We may love to see change in the world or in the lives of others, but never in ours. We all need to have a lifestyle check every now and then. We must see and acknowledge that "the issue of environmental degradation challenges us to examine our lifestyle." (128) The least we can do is to try to rebuild our relationship with God's creation. Pope Francis believes that human conversion requires a proper relationship between humans and nature. Conversion mandates us to acknowledge "our errors, sins, faults and failures" that will enable us to truly repent and decide to make a radical change in our lives. (129)

There is something good every human person can do every day. We are social, political, cultural, and spiritual beings with many gifts, talents, and capabilities. Saints and sages in every generation have proven that Christians can live their lives differently from others. Our capacity for God and others motivates us to live a quality life. The quality of life that people live is often mentioned in the eulogies read at their funerals. What are the lifestyle and quality of Christian life? The spirituality Christians practice, Pope Francis teaches, gives a new perspective on the meaning and standard of human life. Such a life buoys up a far-seeing and ruminative lifestyle that enables people to have the advantage of living a life that is free from consumptive desires and drives. (130)

Seeking a more profound, connected, and active Christian life

Spirituality frees us from all worldly attachments and gives us true joy and peace as it helps us to see the beauty of creation and interact with it regularly. Some people may ridicule us if they see us interact with nature and creation. For instance, some friends of mine used to laugh at me every time they saw my posts of plants, flowers, and animals on my social media page like Facebook. Notwithstanding, many friends love to see my posts and make beautiful comments about them. Those who love the posts, I would conclude, are spiritual people because they can see as St. John Paul II teaches us, that peace is much more than the 'absence of war.' Interior peace, he believed, is very much connected to caring for nature, the environment, and the common good. Practicing Christian spirituality helps us to bring equilibrium into our lives and regain a sense of wonder leading us to a more profound comprehension of life. (131)

Our capacity and practice to appreciate and interact with creation lead us to communion with the Creator. Every created being points to its Creator. Monks and hermits do meditation and contemplation daily in and through nature and creation. Every spiritual exercise helps them to feel, see, and praise God. Even ordinary people may be able to see and appreciate the whole of God's creation and love God more deeply.

Ralph Waldo Emerson wrote:
"The creation of a thousand forests is in one acorn"

Pope Francis inspires us that "when we contemplate with wonder the universe in all its grandeur and beauty, we must praise the whole Trinity." (132). Creation and natural events help us to find God in all things, as St. Ignatius would teach us. Being a Jesuit, Pope Francis has a lot to teach us about the presence of God in His creation. He always invites us to see what God does for the sake of us. Therefore, we must realize that "Christians believing in one God who is Trinitarian Communion suggests that the Trinity has left its mark on all creation." (133) We will find true joy and peace as we continue to take care of creation. Together with all the saints and Pope Francis, we can pray that "our struggles and our concern for this planet never take away the joy of our hope" (134) because we believe that God "has united himself definitively to our earth, and his love constantly impels us to find new ways forward. Praise be to him!" (135).

Edward Schillebbeckx proposes that "human beings must protect nature and guard it against the chaos which human beings can make of it through misbehavior." (136) No matter what, our human behaviors affect nature. Good behaviors protect nature, while bad ones destroy it instantly or eventually. Our human responsibility to care for creation is based on our human conception of it.

Pope Francis, in his encyclical Laudato Si, teaches that 'Judeo-Christian thought' has demystified the natural environment. We still admire its beauty and vastness but have lost our ability to see a divine spark in it. This attitude and mentality emphasize our duty to protect and care for the natural environment. (137) Why must we care for creation? Why care for the earth, specifically? Pope Francis reminds us that we must remember that we are creatures, and the earth that existed long before we were created was given to us as a gift. (138)

We must realize that we are not alone in the universe, but we coexist with other parts of God's creation. How do we relate with the rest of creation which is also a part of God's loving plan? We must realize that basic human needs come from God's creation – food, water, shelter, medicine, and our human needs. For these reasons, we must care for God's creation. In recent years, people have been aware of the ecological crisis. Unfortunately, the beauty of God's creation has been at the disposal of abusive human activities.

Richard J. Woods, in his book Christian Spirituality: God's Presence through the Ages, argues that people from every sector of society, especially religious sectors, have shed light on the need to alleviate the destruction of nature. Spiritual and religious people must have "prophetic and alternative approaches to earth-threatening technologies and globalized economics." (139)

St. Francis of Assisi spoke of Brother Sun and Sister Moon

To bring to our awareness of the need for our care for God's creation, Pope Francis, in his encyclical Laudato Si (On Care for Our Common Home), reminds us that St. Francis would fondly call creatures of God brothers or sisters. A change in his attitude and mentality helped him see other creatures as equals. His relationship with them became more of a fraternal one that quelled exploitative, consumeristic, and limitless use of natural resources and creatures. Overcoming human tendencies to meet all insatiable wants and human's harmonious relationships, St. Francis' simple and poor lifestyle teaches us that created things and beings are not for any reckless and ruthless use but to care for and live with in harmony. (140)

What is our duty or responsibility towards nature and creation?

I often watch the interactions between animals and humans on either YouTube videos or the National Geographic Channel. It is fantastic to see the beauty of interaction. Many owners of dogs always tell us that dogs become aggressive and even vicious to the point of biting strangers when they sense fear in humans. Dogs are also loyal animals who always protect their owners. Speaking of connecting and interacting, many landscapers and owners of gardens remind us that it is a joy to see plants grow and flowers bloom. So, what is our duty or responsibility towards nature and creation?

Thomas Berry concurs that humans must learn to be in communion with the whole of creation to save it from perdition. There is a reciprocity of beneficence between humans and nature. The earth forms one single-family unit where all created realities belong. Each entity has its rights to enjoy, roles to carry out, and dignity and distinctiveness. Each gives voice to the whole of creation for its role in the interconnectedness of all beings and entities on earth and under the heavens. (141)

Learn to look at God's creation with new eyes

There are many reasons why we need to take care of creation, which is one of God's gifts to us. Once creation is taken care of, natural resources shall benefit everyone. It is evident that all people in every generation use God's creation for both their welfare and the community/nation they live in. Conflicts and wars can break out when people have no equal share of natural resources. Some monopolize them while depriving others of their fundamental rights and needs. Some may even try to monopolize both natural and human resources to enhance their political life or booming business.

Thomas L. Leclerc, in his book Introduction to the Prophets: Their Stories, Sayings and Scrolls, explains how Earth's bountiful produce attracts and helps people in politics and business. Thomas contends that the sickle-shaped land, which is extensive, spacious, and outlying, allowed the principal route for trading and traveling for merchants, military forces, and emigrants in the Middle East. (142)

It is not enough to just understand the usefulness of God's creation for our human use. We must also learn to look at it from a new angle. Pope Francis teaches us to change our attitude towards God's creation so that we, like St. Francis, in fidelity to the Word of God, may be able to see God's creation as a splendid book through which our Creator communicates with us and helps us to view His eternal splendor and goodness. Creatures' grandeur and enchantress enable us to know their Creator. God has revealed His everlasting power and divine nature through his works since He created the universe.

St. Francis made sure that a portion of a friary garden was left in its original design to let the natural habitat survive and thrive. Whoever saw natural plants and flowers could contemplate the God who created such beautiful things and find joy and peace in life. (143)

We all grow in every aspect of life, either positively or negatively. For Christians, the most essential thing in life is to grow and mature in values and virtues because it is the only sign of authentic growth. Michael Downey, in his book Understanding Christian Spirituality, posits that "human growth and maturation are prerequisites for growing and maturing in the spiritual life." (144)

As we become mature Christians, we realize that our God is different from other gods because, as Rolheiser proclaims, the God of Christians is both the origin and foundation of all beings with a tangible presence through physical realities, while transcending all of them. We can see, hear, feel, taste, and smell their God because He is the God with the skin on. (145) The concrete presence of God is tangible in and through concrete realities.

Pope Emeritus Benedict XVI, in his book Christianity and the Crisis of Cultures, explains how St. Paul, in his letter to the Romans, sees humans as capable of experiencing God if only they realize that "man is not condemned to remain in uncertainty about God. He can 'see' [Him] if he listens to the voice of God's Being and the voice of [His] creation and lets himself be guided by it." (146) What must Christians do daily then? We can live like saints who see themselves, others, creation, and God differently.

The Blessed Mother and scientists

We all need role models in our lives so that they can inspire, chastise, and guide us on our life's journey. They can be saints, sages, and experts in different fields of study. For example, if we live like the Blessed Mother, we will be able to accept the invitation She offers – to live our ordinary lives in an extraordinary way. It is good to see ordinary daily events through the eyes of holy men and women of our time.

One of the contemporary sages of our times, Basil Pennington, reminds all Christians to live their lives the way the Blessed Mother did because she is:

"the woman, the archetype of the persons we are to be as Christians. He noted that the Blessed Mother summons us to accept and have confidence in life's normality. By living her ordinary life well, she accomplished the august missionary task entrusted to humans by God in human history. She readied herself by living the ordinariness of her life to the fullest, instead of doing any spectacular and extraordinary duties." (147)

The Blessed Mother also came to invite us to redirect our lives to the right path. Her love for us is very real and concrete so much so that we may realize the necessity of taking care of one another better by taking care of God's creation.

Our Lady of La Salette invites us to a conversion of the heart

At La Salette, Grenoble, France, on September 19, 1846, "Our Lady came to warn and to plead. She came to call [people] back, to stir up their faith, to induce them to pray and do penance. Only by prayer and penance can [our] descent into calamity be averted." (148) The Blessed Mother promised that our conversion should help restore everything to its original design that benefits every one of God's beloved children and other creatures. She said, "If [my people] are converted, rocks and stones will turn into heaps of wheat, and potatoes will be self-sown in the land." (149)

Our Lady of La Salette at her Assumption statue on the Holy Mountain

I remember how amazed the first American La Salette Missionaries from Hartford, Connecticut U.S.A., were to see the fertile lands of Myanmar. They remarked, "My gosh, any seed you throw on the ground today grows into a plant/tree the next day." Yes, indeed! I recall how seeds sprout after the first fall of monsoon rain. Typically, we can see the sprouts everywhere the following morning after rainy days.

There is another beautiful side of Myanmar's fertile soil, too. With the change of season come all insects, birds, reptiles, and animals that love fresh new vegetation. In the past (about 80 years ago) my grandparents recalled and told me the truth about Myanmar's beautiful and rich land, "We can see the harmonious relationship between humans and the rest of creation. Everyone used to be patient in waiting for the proper season to harvest any vegetation and hunt for other (than fellow humans, definitely) creatures for food. Things have changed since then – no more patience in waiting for anything.

All of us who prefer quick, healthy, and affordable food wonder how we can maintain our healthy diet daily. We must never forget that to stay healthy is a collaborative and communal effort. It involves everyone and everything God has created because we are all interconnected in many different ways.

The climate clock is ticking, and we're all in this together

We may wonder about our contribution to combatting the climate crisis and the health of God's creation. How much time do we have to care for God's creation? To manage everything wisely and responsibly requires us to be aware of and reclaim our identity as the stewards of God's creation. Benigno Beltran reechoes the warnings scientists have given the world that experts in climate change alert us that we must do something before 2030 to avoid catastrophic damage to the climate and irreparable collapse of the environment worldwide. The constantly rising temperature in the atmosphere (1.5-2.0 degrees Celsius), according the levels before the industrial age will soon cause natural calamities for the earth. Unparalleled torrential floods, drought, destruction of marine life, and scarcity of food will become inescapable realities.

To address these shortages of supplies, nations, and communities must be prepared with systemic programs that are comprehensive and renewal. Only the efficiency, nutritiousness, and healthiness of well-planned programs will save every species, especially humans, from the devasting effects of environmental crises and depletion of natural resources. (150)

We must teach younger generations, as Matthew Fox states in his article: "The Wicked Problem of Climate Change," that the truths we see happening to ecology upskill established religions and belief systems and their principles the virtue of humility. God's creation, like Earth, precedes all religions in existence. Our concern and joint effort to care for creation must unite everyone from all walks of life. (151) There are challenges we must face, problems we must solve, and issues we must resolve daily. Nevertheless, we all are called to be the people who use their reason, knowledge, faith, and love to make our planet "a better place to live in, understanding that Christ's salvation is for all creation." (152)

Our daily responsibility

How shall we live out our Christian responsibility to care for God's creation? There are a lot of simple things we can do on a daily basis:

- ***Food:*** We can cook just enough food to consume for a day or two.
- ***Reduce Waste:*** Reducing wasting resources – food, water, and other necessities – will undoubtedly help us in our every effort to care for God's creation. Those who have much must not waste anything so that those who do not have much or enough will have sufficient resources to live on.
- ***Simplicity of Life:*** Our simple living can help others to simply live their lives. All can also participate in the recycling programs that help reduce pollution in all its forms.
- ***Energy conservation:*** Energy conservation has a big impact on creation, too. If it is sunny and hot, we can dry our clothes in the sunlight. We can also reuse the water that comes from shower rooms (after purifying it) to water plants and other crops.
- ***Restrict pesticides:*** Instead of using pesticides to protect crops, we can use ducks and other animals. Many in Southeast Asia use them for their rice farms. Raising fish in rice fields also benefits rice plants since fish love to eat worms.

The need may be different in developed countries as what they do daily is more of mind work. There has to be a radical change of mindset to help God's creation heal and restore its life to its original design.

Richard C. Schwartz, in his book *No Bad Parts*, raises our awareness of the need for a radical change in what we value and prioritize in life. The time has come for us to listen to, respect, and love Mother Nature more than ever. We should make every effort to have a change of heart and mind for the protection, preservation, and care of our natural environment. (153)

There are many things that people can do who inhabit developed countries. One of them is the proper use of electricity. As to the consumption of electricity, switch it off if no one is using it. It is also a matter of changing our minds for the better and for the common good. Simple little things help us to accomplish complex, great things in life.

St. Therese of Lisieux taught us to execute ordinary chores exceptionally. In seminaries and convents, there are times set for doing daily chores, primarily indoor tasks, as a part of everyday life. Furthermore, everyone has to participate in community work that is outdoors once a week.

St. Thérèse of Lisieux (1873 - 1897)

We all are challenged by the saints and sages of every generation for the proper use of reason and the courageous and faithful practice of our Catholic faith. For instance, "In 1642, at the age of nineteen, [philosopher and physicist, Blaise Pascal] invented an arithmetic machine, the ancestor of our modern computers." Through his invention, Pascal foresaw and proved the reasoning power of the human mind. He urges us to make good use of our reason to comprehend the continuous unfolding realities of life on earth. (154)

All of us who love God and the people of goodwill are always inspired and guided by the wisdom of the Scriptures, holy men, women, and children of every age in our every effort to care for God's creation. For Catholics, the teachings of the Church, theologians, biblical scholars, and spiritual writers are the grammar of our spiritual life, especially in caring for God's creation. Pope Benedict XVI, the grand master of catechesis, as Pope Francis remarked at Pope Benedict's funeral, teaches us the importance of awareness of the connection between God and His creation. Reflecting on the book of Maccabees, Pope Benedict remarked that "during the time of the Exile, the concept of creation becomes central. God is the Creator of heaven and earth." (155)

Artificial intelligence and human responsibility

There was a global outage of the internet on July 19, 2024. It was all over social media and mainline media. Every computer system around the globe that used to operate and guide airlines, banks, hospitals, and stores was affected by it. A priest friend of mine reminded me that humans are not that powerful and not in control of everything. There is always someone who will rise as smarter than the rest of the world. What dawned on me was that every human can do either good or evil, or even both. Geniuses can use their reason and expertise to develop and destroy human communications. On the contrary, humans have God's gifts and talents that can be used and must be used for the common good.

Endowed with inherent gifts from God, humans always find ways and means to advance in their inventions and use of them. Humans can find solutions to all forms of struggles and challenges of life in every age and time. Our ingenuity is spectacular, thanks to divine inspiration and guidance. The Church has continually taught us that we, through our works and endeavors, can strive for and live a life that best suits us.

Nowadays, scientific and technological advancements enable us to master the course of nature. (156) We can think of infrastructures, utilities, medicines, and electronic devices, just to name a few. We can use technological devices for any purpose and at any end we wish. Nevertheless, we Christians also, as people of goodwill, must properly and freely use our natural resources, improve the conditions of our ecosystem, and better ourselves in the process. Even so, we have the duty and responsibility to practice the teachings of Jesus that are focused on serving others. (157)

Can we use artificial intelligence (A.I.) to make our world better?

Pope Francis reminds us that the way we use "artificial intelligence, robotics and other technological innovations" must serve human persons and protect the Earth, our common home. (158) The age of Artificial Intelligence (A.I.) will hopefully be the age of conversion, care for creation and work for the common good.

There are things we all need to do to safeguard our human dignity, creation and the common good. Pope Francis explains that “artificial intelligence ought to serve our best human potential and our highest aspirations, not compete with them.” (159) Socio-political and economic spheres will definitely have different perspectives on and usefulness for AI.

Perhaps faith and common good will have a different worldview than AI. How believers and the people of goodwill prepare themselves for the responsible and proper use of AI is the question we will have to wrestle with and the reality to live with for many years to come.

Questions for reflection:

1. Who is or has been our inspiration and guide in our spiritual and faith life, particularly in the area of conversion and care for God’s creation?
2. What will be our response to the call of God’s creation for our cooperation for its preservation, beautification, survival, and growth?
3. What is your understanding of the Artificial Intelligence (A.I.)? How will you use it effectively in your personal life, while preserving human dignity and practicing your Catholic faith?

Conclusion

We have explored, discussed, and hopefully understood the three essential components of our human and Christian responsibility, namely, 1) conversion, 2) care for God's creation, and 3) the common good. It is essential for all of us to seek and find ways and means for our personal and communal conversion as God continues to inspire, empower, and guide us along the way. Experts in religious, spiritual, socio-political, and environmental fields seek to remind, warn, and equip us with the skills and resources if only we would learn to heed their voices. Human identity and dignity are linked with those of Mother Nature.

God uses every instrument available to heal His creation that has been wounded and needs healing. Through her faithful teachings concerning our human needs for conversion and responsibility to care for our natural environment, the Church is involved in this recovery and healing process. God can heal many illnesses both in us and nature have had with the help of humans and other created realities. How we live and relate with the rest of creation can support and strengthen the health and life of both our ecological and human environments.

We are co-responsible with God in Mother Nature's preservation, protection, and promotion. I concur that we must love and complete our duties both individually and communally as well. The common good is one central responsibility that we, especially Christians, must fulfill in our lifetime. It is our God-given task that we must accomplish as much as possible. The inspiration, direction, and empowerment come from the God we believe in and love. The responsible use of our human reason and faithful practice of our faith can guarantee the health and abundance of our natural resources for our use both in the present and the future.

The gifts and challenges ahead

Scientific and technological advancement is evidence of our ingenuity. However, with progress in technology come risks, advantages, and accountabilities. Let us remember that the rise of Artificial Intelligence (AI) can change and determine the future of human life and its environments. We still need to discover, learn, and apply its usefulness in the areas of 1) conversion, 2) care for creation, and 3) promotion of the common good.

How we will be able to spread the good news of God and the usefulness of our inventions and technology is the question we must still wrestle with as we continue to live and witness to our continuing call to conversion, to our responsibility to care for creation, and to our tireless effort to promoting the common good.

Endnotes

4. Matthew 11:7-19
5. Matthew 11:16-17
6. Matthew Kelly, Life is Messy, 53
7. 1 John 4:7-21
8. Gal 4:1-7
9. Luke 2:1-20
10. Filz, Gretchen "The First Marian Apparition in Church History: Our Lady of the Pillar"
11. Salai Zaw Lwin, A La Salette Spirituality of Forgiveness and Reconciliation for Everyday Living (September 19, 2021)
12. Schlewer, 89
13. M. Basil Pennington, in his book, Mary Today: The Challenging Woman, Model for Maturing Christians, 47
14. Delaney, 2
15. Ibidem, 3
16. Ibidem, 7
17. Pennington, 55
18. Ibidem, 60
19. Patrick J. Hartin, Exploring the Spirituality of the Gospels, 2
20. Hartin, 4
21. Jordan Aumann, Spiritual Theology, 18
22. Philip Sheldrake, 198
23. Nock, 264
24. Ibidem, 265
25. Exodus 23:20
26. John Paul II, Fides et Ratio, 16
27. Hartin, A Spirituality of Faith in Action, 29
28. John Paul II in the opening statement of his encyclical letter Fides et Ratio (Ex 33:18; Ps 27:8-9; 63:2-3; Jn 14:8; 1 Jn 3:2)
29. Thomas Merton: Essential Writings, 32)
30. Matthew 10:29-30
31. Minch, 859
32. Francis, Laudato Si, 159
33. Ibidem, 118
34. Ibidem, 76
35. Minch, 862
36. Edwards, 2
37. Patriarch Bartholomew I and St. John Paul II,

Joint Declaration, 10 June, 2002)

38. Fulton J. Sheen, The World's First Love: Mary, Mother of God, 11
39. Rudolf Bultmann, Theology of the New Testament, 228
40. Scarre, 137
41. Isaiah1:18-19
42. John Paul II, Familiaris Consortio, 9
43. Francis, Apostolic Exhortation, Evangelii Gaudium, (24 November 2013, 182: John Paul II, Post-Synodal Apostolic Exhortation Ecclesia in America, 762
44. 2 Chronicles 7:14
45. Ronald Rolheiser, Wrestling with God, 41
46. Ibidem, 89-90
47. Kathryn James Hermes, 59
48. Genesis 1:1-31
49. John Paul II, Common Declaration on Ecological Ethics, 2002.
50. Groody, 299-300
51. Ephesians 4:5-6
52. Stephen O'Kane, November 26, 2009
53. Jane Blewett, "Social Justice and Creation Spirituality," 16
54. Blewett, 20
55. Denis Edwards, "Earth as God's Creation: The Theology of the Natural World in Pope Francis' Laudato Si,'" 8
56. Francis, Laudato Si, 85
57. Ibidem, 235
58. Pierre Teilhard de Chardin, Christianity and Evolution, 28
59. Delio et alii, 7
60. Hebrews 1:1-2
61. Colossians 1:15-20
62. Hellwig, Jesus the Compassion of God, 110
63. Paul VI, Octogesima Adveniens, 21
64. John Paul II, Redemptoris Missio, 15
65. Ferdinand Nwaigbo, "Pope Francis and the Programme of Ecology in a Time of Change," 216
66. John Paul II, General Audience, 2001, 4
67. Nwaigbo, 216
68. John Paul II, Centisimus Annus (On the Hundredth Year Anniversary of Rerum Novarum), May 1, 1991), 39
69. Nwaibo, 217
70. Francis Laudato Si, 203
71. Ibidem, 206
72. Nwaigbo, 222
73. Ibidem, 230

74. Familiaris Consortio, 21, CCC 1657, Dignitatis Humanae, 5
75. Benedict XVI, Meeting with the Holy Father Benedict with the Clergy of the Diocese of Bolzano-Bressanone, August 8, 2008
76. Francis, Fratelli Tutti, 17; Laudato Si' , 57; AAS 107 (2015), 869
77. Moloney, 123, 124
78. Christopher Vogt, 73
79. Romans 8:18-26
80. Patrick Madrid, Surprised by Truth, 13
81. Joseph Ratzinger, Christianity and the Crisis of Cultures, 66
82. Brendan Byrne, "Ignatius Loyola and John Wesley: Experience and Strategies of Conversion," 61
83. John Paul II, Solicitudo Rei Socialis, 38
84. Cinderella Sequeira, "Laudato Si': Caring Beyond Limits With A Cosmocentric World Vision," 431
85. Francis, General Audience, September 1, 2021
86. Hartin, 6
87. Dei Verbum, Second Vatican Council, 25
88. 2 Timthy 3:16
89. Frank P. DeSiano, The Seven Commandments of Discipleship, 13
90. Ibidem, 15
91. Ibidem, 20
92. Genesis 1:27
93. Konrad Raiser, Challenges and Hopes for a New Millenium to be the Church, Risk Book Series, 36
94. Arthur Darby Nock, 7
95. Frank DeSiano, Why not Consider Becoming a Catholic, 84

96. Marcel Schlewer, 89
97. Ibidem, 91
98. John Paul II, Familiaris Consortio, 9
99. Jeremiah 13:23
100. Thomas Merton, 420
101. Caitlin Morneau, Harm, Healing, and Human Dignity, 16
102. Ibidem, 26
103. Murray, 7
104. Francis, Laudato Si, 217
105. Laudato Si, 221
106. Xue Jiao Zhang, "How St. Francis Influenced Pope Francis' Laudato Si," 48
107. Matthew Fox, 9
108. Nelson Kloosterman, 129
109. Groody, 307

110. Ronald Rolheiser, The Holy Longing: The Search for a Christian Spirituality, 16
111. Rolheiser, The Holy Longing: The Search for a Christian Spirituality, 18
112. Minch, 846

113. Minch, 848
114. Minch, 851, 852
115. Minch, 853
116. Richard Leonard, Where the Hell is God?, 20
117. Minch, 855, 856
118. Leonard, 20
119. Minch, 857
120. Laudato Si, 48
121. Paul VI, Apostolic Letter Octogesima Adveniens (14 May 1971), 21: AAS 63 (1971), 416-417
122. John Paul II, Encyclical Letter Redemptor Hominis (4 March 1979), 15: AAS 71 (1979), 287
123. Benedict XVI, Address to the Diplomatic Corps Accredited to the Holy See (8 January 2007): AAS 99 (2007), 73
124. Benedict XVI, Encyclical Letter Caritas in Veritate (29 June 2009), 51: AAS 101 (2009), 687
125. Laudato Si, 13
126. Ibidem, 164
127. Ibidem, 189
128. Ibidem
129. Ibidem, 206
130. Benedict XVI, Jan. 1, 2010, Message for the 2010 World Day of Peace, 11: AAS 102 (2010), 48
131. Laudato Si, 218
132. Ibidem, 222
133. Ibidem, 225
134. John Paul II, Catechesis (2 August 2000), 4: Insegnamenti 23/2 (2000), 112
135. Laudato Si, 239

136. Ibidem, 244
137. Ibidem, 245
138. Schillebeeckx, Church, 242 [244]
139. Laudato Si, 78
140. Ibidem, 67
141. Richard J. Woods, Christian Spirituality: God's Presence through the Ages, 258
142. Laudato Si, 11
143. Thomas Berry, The

Great Work: Our Way into the Future, 4
144. Thomas L. Leclerc, Introduction to the Prophets: Their Stories, Sayings and Scrolls, 2-3
145. Laudato Si, 12
146. Michael Downey, Understanding Christian Spirituality, 8
147. Rolheiser, 81
148. Benedict XVI, Christianity and the Crisis of Cultures, 96
149. Pennington, 120
150. Delaney, 91
151. Ibidem, 76
152. Benigno Beltran, 54
153. Matthew Fox, "The Wicked Problem of Climate Change", 588
154. Sequeira, 433
155. Schwartz, No Bad Parts, 67
156. Francis, Apostolic Letter Sublimitas Et Miseria Hominis
157. Benedict XVI, What is Christianity: The Last Writings, 41-42
158. Francis, World Economic Forum, January 22, 2018
159. Francis, World Day of Peace, January 1, 2024

References

Primary Sources:

Benedict XVI, Pope, *Encyclical Letter Fides et Ratio (On the Relationship between Faith and Reason).* Holy See. https://www.vatican.va/content/john-paul-ii/en/encyclicals/documents/hf_jp-ii_enc_14091998_fides-et-ratio.html.

____________________. *What is Christianity: The Last Writings.* Elio Guerriero and Georg Ganswein, eds. and Michael J. Miller, trans. San Francisco, CA: Ignatius Press, 2023.

Francis, Pope, Apostolic Letter *Sublimitas Et Miseria Hominis (On The Fourth Centenary of the Birth of Blaise Pascal*), Italy, Rome: John Lateran, 19 June 2023).

_____________. "Message of His Holiness Pope Francis to The Executive Chairman of the "World Economic Forum on the Occasion of the Annual Gathering in Davos-Klosters, 23-23 January 2018).

____________. "Message of His Holiness Pope Francis for the 57th World Day of Peace, January 1, 2024. https://www.vatican.va/content/francesco/en/messages/peace/documents/20231208-messaggio-57giornatamondiale-pace2024.html.

John Paul II, Pope, Common Declaration on Environmental Ethics (10 June, 2002). The Holy See.

______________. General Audience (January 17, 2001). Holy See.

______________. Encyclical Letter *Solicitudo Rei Socilais (On Social Concerns*), 30 December 1987. Holy See. https://www.vatican.va/content/john-paul-ii/en/encyclicals/documents/hf_jp-ii_enc_30121987_sollicitudo-rei-socialis.html.

Paul VI, Pope, Apostolic Letter *Octogesima Adveniens (Eightieth Anniversary of Rerum Novarum)*, 14 May, 1971. Holy See.

______________. Pastoral Constitution *Gaudium Et Spes (The Church in the Modern World)*, 7 December, 1965). Holy See.

The United States Conference of Catholic Bishops (USCCB). *The Compendium on the New Evangelization: Texts of the Pontifical and Conciliar Magisterium.* Washington, D.C: The United States Conference of Catholic Bishops, August, 2015.

Secondary Sources:

Delaney, John J., *A Woman Clothed With The Sun: Eight Great Apparitions of Our Lady.* John J, Delaney, ed., New York, NY: Image Books/Doubleday, Random House, Inc., 1960.

Beltran, Benigno P. "Erath Stewardship, Economic Justice, and World Mission: The Teachings of Laudato Si." In *Missiology: An International Review,* 2020, Vol. 48(1)39-56, doi: 10.1177/0091829619897432.

Berry, Thomas. The Great Work: Our Way into the Future. Bell Tower: 1999.

Blewett, Jane. "Social Justice and Creation Spirituality." *The American Theological Library Association,* 2016.

Bultmann, Rudolf. *Theology of the New Testament.* Kendrick Grobel, trans. Waco, TX: Baylor University Press, 2007.

Byrne, Brendan. "Ignatius Loyola and John Wesley: Experience and Strategies of Conversion. *Colloquium 19* (1), 54-66, 1986.

Chardin, Pierre Teilhard de. *Christianity and Evolution.* Orlando, FL: Harcourt, Inc., 1969.

Daniel G. Groody. "Homeward Bound: A Theology of Migration for Fullness of Life, Justice and Peace." *The Ecumenical Review.* Volume 64, Number 3, October 2012.

Delaney, John J., ed. *A Woman Clothed With The Sun: Eight Great Apparitions of Our Lady.* John J, Delaney, ed., New York, NY: Image Books/Doubleday, Random House, Inc., 1960.

Delio, Ilia, Warner, Keith Douglass and Wood, Pamela. *Care for Creation [a Franciscan spirituality of the earth].* Cincinnati, Ohio, 1999.

DeSiano, Franck P. *The Seven Commandments of Discipleship: What God Asks of Us.* Mahwah, N.J: Paulist Press, Inc., 2003.

___________________. *Why Not Consider Becoming A Catholic.* Washington, D.C: The Paulist National Catholic Evangelization Association, 2010.

Downey, Michael. *Understanding Christian Spirituality*. Mawah, N.J., Paulist Press, 1997.

Edwards, Denis. "Earth as God's Creation: The Theology of the Natural World In Pope Francis' Laudato Si'."

Filz, Gretchen. "The First Marian Apparition in Church History: Our Lady of the Pillar" (October 11, 2016).

Fox, Matthew. "Creation Spirituality: A Personal Retrospective," *Listening* 24 (Spring 1989).

______________. "The Wicked Problem of Climate Change: Climate Change, Laudato Si', Creation Spirituality, and the Nobility of the Scientist's Vocation." *Zygon*, vol. 53, no. 2 (June 2018)

Francis, Pope. "Pope's September Prayer Intention: An Environmentally Sustainable Lifestyle." The General Audience. September 1, 2021.

Hartin, Patrick J. *Exploring the Spirituality of the Gospel.* Collegeville, Minnesota: Liturgical Press, 2011.

Hellwig, Monica. Jesus, *The Compassion of God: New Perspectives on the Tradition of Christianity,* Theology and Life Series 9. Wilmington, Delaware: Michael Glazier, Inc., 1985.

Hermes, Kathryn J. *15 Steps to Spiritual Healing: Making Peace with Yourself.* Boston, MA: Pauline Books & Media, 2007.

Kelly, Matthew. *Life is Messy.* North Palm Beach, FL: Blue Sparrow, 2021.

Kloosterman, Nelson D. "Environment as Religion: Matthew Fox's Creation Spirituality as a Paradigm for Environmental Ethics." *Mid-America Journal of Theology* (MJT 9/1, 1993), 124-141.

Leonard, Richard. *Where the Hell is God?* (Mahwah, NJ: HiddenSpring, Macarthur Boulevard, 2010).

Leclerc, Thomas L. Introduction to the Prophets: Their Stories, Sayings and Scrolls (Paulist Press, Mahwah, NJ, 2007).

Madrid, Patrick. Surprised by Truth: Eleven Converts Give the Biblical and

Historical Reasons for Becoming Catholic. Encinitas, CA: Basilica Press, 1994.

Merton, Thomas. Thomas Merton: Essential Writings (Selected with an Introduction by Christine M. Bochen). Maryknoll, New York: Orbis Books, 2000.

Minch, Daniel. "Our Faith in Creation: God's Faith in Humanity: Edward Schillebeeckx and Pope Francis on Human Transcendence and an Anthropocentric Cosmos."

Morneau, Caitlin. Harm, Healing, and Human Dignity: A Catholic Encounter with Restorative Justice. Collegeville, Minnesota: Liturgical Press, 2019.

_________________. *Poems to Live By In Uncertain Times.* Edited by Murray, Joan. Boston, MA: Beacon Press, 2001.

Nock, A. D. Conversion: The Old and the New in Religion from Alexzander the Great to Augustine of Hippo (The Johns Hopkins University Press, Baltimore, Maryland, 1998).

Nwaigbo, Ferdinand. "Pope Francis and the Programme of Ecology in a Time of Change."

Pennington, M. Basil. *Mary Today: The Challenging Woman, Model for Maturing Christians.* Doubleday, New York: An Image Book, 1987.

Raiser, Konrad. *Challenges and Hopes for a New Millennium to be the Church.* Geneva: WCC Publications, Risk Book Series, 1997.

Ratzinger, Joseph. *Christianity and the Crisis of Cultures.* Marcelolo Pera, trans. Ignatius Press, San Francisco, 2006.

Moloney, Raymond. "Conversion and Spirituality: Bernard Lonergan (1904-1984)." *The Way*, 43:4 (October 2004), 123-134).

Rolheiser, Ronald. *The Holy Longing: The Search for a Christian Spirituality.* IMAGE and Random House Company, NY, 1999.

_________________. Wrestling with God: Finding Hope and Meaning in Our Daily Struggles to Be Human. Penguin Random House LL.C, NY, 2018.

Scarre, Geoffrey. "Political Reconciliation, Forgiveness and Grace." Studies in *Christina Ethics*, 24(2), 171-182, 2011, doi:10.1177/0953946810397442.

Schillebeeckx, Edward. "Church and World," in *World and Church, Collected Works of Edward Schillebeeckx*, 4 (London: Bloomsbury, 2014, 83 [110].

Schlewer, Marcel. *All My People: Why She Spoke, Why She Wept at La Salette* (Grassroots Publishing International, Enfield, NH, September 19, 1998).

Sheldrake, Philip. *Spirituality and History: Questions of Interpretation and Method.* New Edition. Maryknoll, NY: Orbis Books, 1995.

Sequeira, Cinderella. "Laudato Si': Caring Beyond Limits With A Cosmocentric World Vision." *Theology*, 2019, Vol 122(6)427-434, doi: 10.1177/0040571X19872106.

Vogt, Christopher. "Ecological Conversion, Spirituality, and Solidarity – Selected Session." *CTSA Proceedings*, 72, 2017.

Woods, Richard J. *Christian Spirituality: God's Presence through the Ages.* New Expanded Edition. Maryknoll, NY: Orbis Books, 2006.

Zhang, Xue Jiao. "How St. Francis Influenced Pope Francis' Laudato Si." *Crosscurrents*, Vol. 66, No. 1, Religion and Climate Change, University of North Carolina Press, 2016.

Further Readings:

Minch, Daniel. "Our Faith in Creation: God's Faith in Humanity: Edward Schillebeeckx and Pope Francis on Human Transcendence and an Anthropocentric Cosmos."

Nwaigbo, Ferdinand. "Pope Francis and the Programme of Ecology in a Time of Change."

Edwards, Denis. "Earth as God's Creation: The Theology of the Natural World In Pope Francis' Laudato Si'."

Shrine Church on the Holy Mountain of La Salette in southeastern France; inset: On the Holy Mountain, the bronze statue of Our Lady of La Salette with the special crucifix with its hammer and pincers.

Made in United States
Orlando, FL
10 November 2026

90498882R00066